COSMIC
CHRISTIANITY

THE CHANGING
COUNTENANCE
OF COSMOLOGY

COSMIC CHRISTIANITY

THE CHANGING COUNTENANCE OF COSMOLOGY

Willi Sucher

✑ Anthroposophic Press

The Changing Countenance of Cosmology and
*Cosmic Christianity: The Stars During the Three Years of Christ's Ministry
and Practical Viewpoints with Regard to Evolution* were originally printed
and published at the Astrosophy Research Center, Inc.
P.O. Box 13, Meadow Vista, CA 95722
Copyright Astrosophy Research Center, Inc., 1985.

This edition published in the United States by the Anthroposophic Press,
R.R. 4, Box 94 A-1, Hudson, New York 12534.

Library of Congress Cataloging-in-Publication Data
Sucher, Willi, 1902–1985.
[Cosmic Christianity]
Cosmic Christianity: and, The changing countenance of cosmology
/ by Willi Sucher.
First work originally published: 1970.
Second work originally published: 1971.
Includes bibliographical references.
ISBN 0-88010-369-8 (paper : alk. paper)
1. Anthroposophy.   2. Astrology.   3. Jesus Christ—Anthroposophical
interpretations.   I. Title: Cosmic Christianity.   II. Title: Changing countenance
of cosmology.
BP596.Z6S83   1993
299'.935—dc20   93-201
CIP

Cover design: Barbara Richey

10 9 8 7 6 5 4 3 2 1

Printed in the United States of America

CONTENTS

PART TWO

COSMIC CHRISTIANITY

*The Stars during the Three Years of Christ's Ministry
and Practical Viewpoints with Regard to Evolution*

---

*The diagrams in this book are
the original work of Willi Sucher.*

FOREWORD

It is with joy and gratitude that I welcome this first printed edition of Willi Sucher's works, *The Changing Countenance of Cosmology* and *Cosmic Christianity*. Born at the beginning of this century, Willi Sucher devoted his life to helping humanity find a renewed relationship to the world of the stars. Working out of the suggestions of Rudolf Steiner and his close colleague Elisabeth Vreede, Sucher laid an all-embracing foundation for a new Christian star wisdom within which we human beings are shown our potential to rise to co-creators within the ongoing stream of evolution.

This publication presents the contents of a series of lectures and "work" sessions in which there was an active exchange between lecturer and participants at Hawkwood College, England, in 1969. Soon after, Willi Sucher revised them for publication in booklet form. For the current publication, further editing and minor revisions have been made, still preserving the individual style of presentation of the original text.

A short biographical introduction to Willi Sucher follows, giving the background of the unfolding of his life work, which has become known as Astrosophy, Wisdom of the Stars. As can be read there, shortly before his death in 1985, he founded the Astrosophy Research Center with a group of supporters. This center now houses and cares for his copious notes, charts, personal library, and various leaflets on his research work, all of which are available for study at the center. There are also a lending library, various leaflets for sale, and the possibility for

accommodations on request. For further information, contact
the Astrosophy Research Center, PO Box 13, Meadow Vista, CA
95722, USA; (916) 878-2673.

May this be the first of other publications that will enable
more people to become aware of Willi Sucher's gift to humanity,
that they may join those of us who are already grateful for the
enrichment he has brought to our lives. Born out of spiritual
science, this new star wisdom is already helping in many
realms of life, for instance in agriculture, medicine, and education. It bears a potential by which we can develop an insight to
enable us to meet the challenge of our time in a positive way.

HAZEL STRAKER

27 February 1992

BIOGRAPHICAL INTRODUCTION

Willi Sucher (1902–1985)

FOR THOSE READERS new to the work of Willi Sucher, we begin this book with a short biography as an introduction to the life and work of this pioneer in the spiritual science of astrosophy. The content of the books assumes a familiarity with the work of Rudolf Steiner, for it is entirely on the foundation of Anthroposophy that Willi Sucher's research is based. Therefore, a study of the basic works of Steiner is recommended in order to fully enter into the content presented here.

At the beginning of this century, Rudolf Steiner began to speak of a new way of knowing spiritual realities, which he called Anthroposophy, wisdom reborn through the human being. Until his death in 1925, he brought to humanity the means for renewal through the cultivation of a modern spiritual science applied to a wide range of practical spheres of life, including education, medicine, agriculture, the arts, and social forms. Behind these areas of applied Anthroposophy stands an all-embracing wisdom of the relation of the human being to the cosmic world in the past, present, and future. Many individuals experienced Rudolf Steiner and were inspired to take up one area to develop as a life task. One of these individuals was Willi Sucher, who as a young man heard Steiner speak. Sucher was especially inspired by Steiner's vision of the new growing relationship of the human being to the cosmic world, that of co-creator with the divine beings. Steiner called for this relationship to become more and more of a reality, and Sucher picked up the challenge and devoted his life to the task of developing a new wisdom of the stars, astrosophy, that would recognize this new role of the human being in earth evolution.

Willi Sucher was born on August 21, 1902, in the southern German town of Karlsruhe to a young bookkeeper and his wife. His earliest memories were of the death of his mother when he was four. This great loss marked the beginning of a childhood in which he experienced himself as the unwanted stepchild. His father soon remarried, and Willi's new stepmother was, as he would later describe her, "just fiercely against my very existence." He was rejected and mistreated during these years, until at age thirteen he was sent to live with his uncle's family when his father was called up to serve in the military during World War I. Though his uncle, Karl Sucher, was kind, Willi was an extra mouth to feed in a large family that was suffering under the economic hardships of the war, and so after a year he was sent to live with relatives of his mother on a small farm in a nearby town. Life there was very strenuous. The family worked late into the night in the fields, so Willi was responsible for tending the house and cooking for the others after a full day at school. He would later characterize his childhood as always being "just one too many," but he saw in this a positive force in his life that served to build inner strength and perseverance in the face of hardship.

In 1918, at the age of sixteen, Willi decided to make himself independent. He saw that his original hope of becoming an architect was impossible due to his financial circumstances, so he applied to become an apprentice in a bank. He was accepted and began the two-and-a-half-year training. He did not like it, but he would continue in this profession for twenty-one years. He often pointed out how through this work he learned rigorous attention to details and accuracy in calculation. This mathematical training would serve him well later in his real life's work.

In 1919, Willi came into contact with the ideas of Rudolf Steiner. His uncle Karl, with whom he had kept in touch, had heard Rudolf Steiner lecture and spoke to Willi about him. Willi recognized immediately that these ideas would become his path in life and soon applied for membership in the

Anthroposophical Society. However, he was told he must wait another year until he was eighteen. His uncle Karl also spoke with him about astrology, expressing his concern about its unsuitability for modern humanity. Willi recalled one such conversation in which his uncle spoke of how important it would be that someday an anthroposophist bring new light to the whole field of astrology through the insights of spiritual science. His response was, "Why should we wait? Can't we do it ourselves?" He was eighteen years old, the time of his first moon node return. He then began to read the literature on astrology in an effort to understand it, only to turn away from it time and again, repelled by its determinism, which he felt degraded the true dignity of the human being.

He continued also his study of Anthroposophy, attending lectures when possible and reading. In 1922, he became inspired by Steiner's ideas on the Threefold Social Order and moved to Stuttgart to join a small bank, *Bankhaus Der Kommende Tag,* which was connected with several businesses trying to put these ideas into practice. On one occasion Rudolf Steiner visited the bank and was introduced to all who were working there. Willi was deeply impressed by this personal encounter and with the way Steiner so fully entered into their situation. It was typical of Willi that when asked whether he had ever requested a private interview with Rudolf Steiner, he replied that he had never felt his personal questions should take up the precious time of such a busy man. However, due to the increasing economic difficulties of those times, the bank was forced to close. Willi then took a position in a bank in Bruchsal. Through a friend he met his future wife, Helen, who lived with her parents in Stuttgart and was also attending the lectures of Rudolf Steiner. They both joined the newly formed Christian Community and were married in 1927 by Dr. Friedrich Rittelmeyer, the founder of the Christian Community and a leading Lutheran theologian in Germany at the time. Willi was working and living in Bruchsal with some anthroposophists, and Helen was living with her parents in Stuttgart, so Willi would visit on weekends by train. They were soon able to get their own place and in 1927 moved into a small two-room apartment with a kitchen but no bathroom. Willi would return home from his work at the bank and put in two hours of study in the evenings. On weekends for recreation they would go for hikes in the Black Forest.

The year 1927 was a significant year in Willi's life, not only due to his marriage, but for another reason as well. He came across the report of a lecture given by Dr. Elisabeth Vreede, the head of the Mathematical-Astronomical Section of the School for Spiritual Science at the Goetheanum in Dornach, Switzerland, where the Anthroposophical Society was centered. (Dr. Vreede had first met Rudolf Steiner in 1903, when she was a member of the Theosophical Society. She became one of the group of individuals who worked with Steiner in developing the Anthroposophical Society, moving with him from Berlin to Dornach in 1917 to help build the first Goetheanum. She was an original member of the Vorstand of the Society and was appointed head of the Mathematical-Astronomical Section by Steiner.) In this lecture Dr. Vreede referred to remarks by Rudolf Steiner about the configurations of the heavens at the time of the passing over of a human being into the spiritual world at death. Looking back at this moment forty-two years later, Willi wrote:

> This picture struck home like lightning. Here arose a perspective which no longer depicted the human being as a helpless object of the rhythms and movements of the stars. It was the soul of man which meant something for the stars; they were even waiting for that which he had to bring them as the fruits of his earth-experiences. A ray of hope which seemed to shed light on man's quest for spiritual freedom fell on the complex of astrology.
>
> Subsequent researches—just on the basis of the mathematics of planetary rhythms—fully confirmed that hope. Indeed, the biographical rhythms of a great number of historic personalities proved to coincide perfectly with the configurations of the heavens at the moment of their passing over. *The experience that man was not only a creature but was on the road to becoming a co-operator even with the cosmos shaped itself increasingly.* This gave hope that similar constructive views might eventually be found with

regard to man's association with the stars at the moment of his incarnation. Later discoveries proved that this was no vain hope.

Now his studies took on an ever-deepening intensity. He was at that time studying the biography of Tolstoy. He worked out the configurations of the heavens at the time of Tolstoy's death and after careful deliberations sent this star picture, along with some very tentative suggestions, to Dr. Vreede in Dornach. She responded, as Willi would later say, "very positively" and invited him to Dornach the week after Easter, 1928. Willi was then twenty-five years old. Dr. Vreede, as part of her task as leader of the Mathematical-Astronomical Section at that time, was giving lectures and courses, and between 1927 and 1930 published forty-two letters on the theme "Astronomy and Anthroposophy" (revised and published in book form in 1980 by the Philosophisch-Anthroposophischer Verlag at the Goetheanum). Her research found an eager pupil in Willi, and this became the starting point for a working relationship that would develop over the next ten years, during which Dr. Vreede gave Willi encouragement and challenges to further develop the work she had begun. Willi often described how Dr. Vreede would send him a statement by Steiner on something about the relation of the human being to the cosmos with the command, "I cannot do it. You must do it!" Willi would then work out astronomically/mathematically his understanding of these indications and send them back to Dornach. As he later wrote, "It was Dr. Elisabeth Vreede who suggested that I investigate the connections of the human being with the prenatal star-events, i.e., during the embryonic development. She advised me to employ for this purpose the ancient 'Hermetic Rule'—originating in Ancient Egypt.

Willi often traveled to Dornach during these years and in 1931, at twenty-nine years of age, he was invited to lecture at the Goetheanum and later at the Anthroposophical Clinic in Arlesheim. In 1934–35, Dr. Vreede, on behalf of the Mathematical-Astronomical Section, published a series of *Astrologische Betrachtungen* ("Astrological Studies") written by Willi, except for the first one in which she wrote:

The following studies are meant to inform the reader about the investigations of our co-worker Willi Sucher, as he has developed them in conjunction with the Mathematical-Astronomical Section for some years now. Willi Sucher's point of departure has not been traditional astrology— which was known to him—but Rudolf Steiner's spiritual science, especially Rudolf Steiner's suggestions concerning the realm of astrology.

As Willi's work continued, conditions in Germany were becoming increasingly difficult with Hitler's rise to power. Willi and his wife realized they would not be able to continue their work in this environment. His correspondence with Dr. Vreede had been intercepted, and astrologers in Germany were being arrested. Anthroposophists had to meet secretly in small groups. Willi later spoke of the need for secrecy during these times and described how he and Helen would go into the kitchen, fill the sink, and place a pot over the drain, in order to speak about things that should not be overheard by neighbors who might report them. In 1936, Willi was again in Dornach visiting Dr. Vreede, after her expulsion from the Vorstand along with her colleague Ita Wegman and other original members. It was through her efforts after this visit that he was invited in 1937 to lecture in Holland and then in England at the Rudolf Steiner House in London, where George Adams translated his lectures. During this visit to England, Willi and Helen actively looked for opportunities to leave Germany. He spoke with his friend Eugen Kolisko, who introduced him to Fried Geuter, the co-founder of Sunfield Home, an anthroposophical home for handicapped children at Clent. Geuter said, "Mr. Sucher, come to us and I shall build you an observatory!" So on their return to Germany, the application process for a visa was begun. Their intention to emigrate had to remain a secret, except for a few close friends within the Anthroposophical Society. After several months, just after Easter 1938, the necessary papers were obtained, and the Suchers left their homeland for England on what was officially considered a "visit," with only twenty marks and a few personal belongings. They would not return to live in Germany again. Willi was thirty-five years old.

So after twenty years of working in banks, Willi was plunged into working in a nursery with severely handicapped children, while Helen worked in the kitchen. The language was also new. Fried Geuter exhorted his teacher there in the nursery, "Teach him in English, but scold him in German!" Here Willi gathered the clinical experience which he would later unite with Steiner's indications—to develop the idea that a dedicated staff, working with a deep knowledge of a child's star configuration, could effect healing. In describing such work, he said, "Often we would work deep into the night, and the next day the child was a different being." Of this time he also said, "So you see, it really was an 'observatory'...of the cosmic influences on human destiny!"

In 1938, Willi was able to meet once again with Dr. Vreede at a conference held in Bangor, Wales, near Penmaenmawr. At this conference, Valentin Tomberg lectured on the working of the spiritual hierarchies in the twentieth century. Willi described how he and Dr. Vreede climbed a hill behind Penmaenmawr to two Druid stone circles: "So we took leave of one another at least for the time being, in the proximity of witnesses to an age-old star wisdom and with a deep feeling of responsibility for its future." This was the last time Willi saw Dr. Vreede. Her last years were lonely ones. On account of the war she was cut off from her friends in Holland, England, and Germany. The death of Ita Wegman, her close friend and colleague, in 1943 came as a great shock. Just two months later she fell ill and moved to southern Switzerland in the hope that the warmer climate there would help improve her condition. But this was to no avail, and at 4:45 p.m. on the afternoon of August 31, 1943, she breathed her last breath, having lived a rich life dedicated to Rudolf Steiner and Anthroposophy.

Due to the war there was much concern in Great Britain that there were enemy informers among the many refugees. Thus, all German and Austrian men and some women were rounded up as "enemy aliens" to be interned in detention camps around the British Isles until their motives could be investigated. Anticipating this, Willi packed a small case with his most precious tables for working out star positions and other aids he needed to carry on his research. In those days modern ephemerides, computers, and calculators were not available; all had to be worked out using special tables that were not easily obtainable. Thus when the police came for him and some others working at the home, he was fully prepared. Before they were taken away, all of the co-workers stood in a big circle and sang a song that had become a leading motif for the aim of their work with the children, "In the Quest of the Holy Grail," to bid them farewell.

They were first taken to a kind of clearing house at a military barracks at Worcester, then on to a place near Liverpool. While his personal belongings were being searched for possible subversive material, the first man in their group told the inspectors that they had come from a home for children that had had an outbreak of scarlet fever, which was true. Because of this, their group was taken to an isolation unit, and their belongings were sent with them without being inspected. Thus Willi's research materials were spared. The group was later transferred to a camp on the Isle of Man, in the Irish Sea, formerly a holiday resort with various hotels that had been taken over and surrounded with barbed wire. Here several anthroposophists found themselves interned together—Dr. Ernst Lehrs and Dr. Karl König, the founder of Camphill, among others. Willi would later describe this time as a most fruitful period of research. For eighteen months, these individuals were given time to hold a kind of "super college," as Willi called it. They had long conversations about their studies and research, and practiced giving talks to each other. All of their material needs were provided, and they were left free to organize their daily lives within the compound. They could go out and work on farms, which Willi tried but found too strenuous. Since he was a medical doctor, Dr. König was released first, in January 1942, and he secured Willi's release by inviting him to come to work at his home for children in Aberdeen, Scotland. Willi was joined by Helen, who had remained at Sunfield, and he carried on his researches while helping to care for the children. He also started to write a "Monthly Letter" for a number of subscribers and to give lectures. It was at one of these lectures, given on a return visit to Sunfield Home in Clent, that Hazel Straker first met him. She describes this meeting in the following way:

I have very vivid memories of him coming to lecture at Sunfield where I had come to work just after his internment. I remember this modest man, immaculately dressed, walking up and down in front of us, telling in a sure but quiet and pictorial way about his researches into the gestures of the stars during the three years that Christ worked on earth. It was about the gestures of Mercury, its meetings with the Sun, and their relationship to the seven signs or miracles described in the Gospel of St John. The pictures he painted in the air have remained with me, growing as a reality that means much to me still, having become intimately related to daily life. On another occasion he spoke of the cycle of the year, and from that I carry the certainty of the "living being of the Sun." The depth of his disciplined research work shone through, radiating confidence, which left one totally free.

Later, in 1944, when Willi and Helen returned to Sunfield, Hazel Straker came to work more and more closely with him and his research. In 1946, at the request of Eleanor Merry and Maria Schindler, Willi and Helen moved to London for a short time and taught evening classes. In 1947, Dr. Alfred Heidenreich, the founder of the Christian Community in Britain, instigated an invitation to Garvald, a curative home in Scotland, where Willi became the director for a short time. Here Hazel Straker joined them as a co-worker. At Garvald Willi came into conflict with one of the co-workers on account of his "astrological" work with the children, so they left when Dr. Heidenreich invited them to work at Albrighton Hall, a center for Christian Community conferences, near Shrewsbury. This time, Willi wrote, "was one of the most positive and creative periods of my life. Dr. Heidenreich gave me absolute freedom to develop my work." Here the English manuscript of *Isis Sophia*, published in 1951, was prepared (it had already been published in Germany), as well as *Man and the Stars*, the second Isis Sophia series, published in 1952. Also during these years the "family" of co-workers was formed. Willi and Helen had no children, but co-workers came who would give lifelong support to the work. They joined together in research and practical life. In addition to Hazel Straker, Helen Veronica Moyer and her sister, the artist Maria Schindler, came together in this work. They cared for the conference house and assisted in the star work, allowing Willi time for research besides lecturing at conferences there and traveling to meet increasing requests to speak to other groups in England, Scotland, and Holland. During this time the work on the starry background of the Greek, Norse, and Celtic mythologies was done, much research into historic periods and personalities was conducted, and the very new areas of heliocentric and lemniscatory views of the universe were explored.

In 1953, the conference house closed for financial reasons, and the group moved to Larkfield Hall, a curative home in Kent, England, where they were able to build a small house through the help of a devoted friend. Because of his lecturing commitments, Willi did not have time to work with the children, but his co-workers did, and they would sit together with Willi and work over the children's incarnation charts. This work was described by Hazel Straker:

> This was not just a horoscope, the stars at the moment of birth, but a picture of the gestures of the stars during the nine months' preparation for birth, the embryonic development. Dr. Vreede had introduced this, and following her request, Willi had done much further research. This meticulous, painstaking work that he had carried out over the last years showed rich fruits as he led us through the starry events to the great imaginations behind, which were able to inspire us in a very helpful way for our further work with the individual children. Here too the recurrence of gestures connected with the deeds of Christ during the three years that He worked on earth were an integral part of our considerations. Although I had already committed myself to this work, it became ever clearer that here was a great potential for true healing.

In 1955, Willi was invited to America to lecture at the Three-fold Farm anthroposophical community in Spring Valley, New York. During this first trip to the U.S., he gave seventy lectures or workshops in his nineteen-week stay, which included a visit to

Los Angeles to teach a course in the teacher training program at the Highland Hall Waldorf School. This California connection would play a significant role in his later destiny.

On returning to England, Willi began to work on the book *Drama of the Universe.* The two previous books, *Isis Sophia* and *Man and the Stars,* had been written from the geocentric (Earth-centered) perspective. Now his researches into the heliocentric perspective had progressed to the stage of putting them into this book. It was a big task, with much of the work of preparing it for publication being done by Hazel Straker and Veronica Moyer. It was published in 1958, and to celebrate they all decided to take a holiday. Helen had always wanted to see palm trees, so they decided to make a journey to Egypt. But as plans were being finalized, Willi suggested, "Why go east, why not go west to America?" So the family of co-workers journeyed across America, from Montreal to Denver, through Salt Lake City, on to Los Angeles. Here Helen saw her palm trees, and it was here that she decided they would stay. The persistent requests to come and join the work at the Highland Hall School and to begin a much-needed school for curative education were another reason to move to California, so they decided to immigrate. They returned to England, sold their home, and in 1961, through the generosity of a friend, this group of four founded the Landvidi Center for Exceptional Children in Los Angeles, which operated under their guidance for seven years. During these years Willi gave many lectures and courses in other parts of the States, as well as returning to Europe to lecture in England, Holland, Switzerland, and Germany.

With the closing of the school in 1968, the Suchers searched for an area to retire to. Many places were considered, including some in England and Canada, but eventually their choice was Meadow Vista, a small town on the lower slopes of the Sierra Mountains not far from Sacramento. Now there was more time to devote to writing and research. Willi also continued his traveling lecture activities, besides holding courses and study groups in his home. During this time an increasing stream of individuals came seeking help in their lives, and Willi's work with the profound pictures given in the birth and prenatal asterograms brought light onto the destiny path of those who sought him out. It was during this time that *Cosmic Christianity* (1970) and *The Changing Countenance of Cosmology* (1971) were published. Both of these books, of which this volume is composed, are the content of a series of parallel morning and evening workshops taught by Willi in August 1969 at Hawkwood College in England, which he later wrote down for publication. He later wrote about the research published in *Cosmic Christianity*:

> Finally, I must mention the research work which I did about the Christ Events. I came more and more to the impression that these Cosmic perspectives of the Christ Events are a foundation for the experience of the workings of the Christ Impulse in times after the so-called Mystery of Golgotha. It turned out that whenever one of the cosmic events during Christ's ministry repeats itself, then there is offered the opportunity to understand and even to realize in an inner spiritual sense the significance of the corresponding Deed of Christ. As I said, these possibilities are "offered" to the human being. He can freely accept them and identify eventually with them.

He also continued to write the *Monthly Star Journals* (1965–75) to subscribers. In the November 1970 letter he wrote of his life work since first reading the article by Dr. Vreede:

> It is now forty-two years since this lightning-storm happened. And ever since I have been enabled to carry on this research. Sometimes external circumstances were difficult, but there seemed always to be a helping hand in the background, which often arranged things forcefully in order to facilitate the work.
> As I said before, the road was never easy. Suspicion and distrust acted as forceful breaks. One can fully understand this if one views the grave dangers which beset the road right and left towards a new, constructive astrology. Human egotism is all too easily inclined to misuse this knowledge in ignorance and dilettantism. All throughout the years the shining beacon of Rudolf Steiner's wisdom

was an unceasing encouragement and also consolation when distrust led to direct attack. There is, particularly, one passage in Rudolf Steiner's lecture cycle *Christ and the Spiritual World* (28 Dec. 1913 to 2 Jan. 1914) which I should like to quote: "It became clearer and clearer to me—as the outcome of many years of research—that in our epoch there is really something like a resurrection of the astrology of the third epoch (the Egypto-Chaldean civilizations), but permeated with the Christ Impulse. Today we must search among the stars in a way different from the old ways, but the stellar script must once more become something that speaks to us." (Lecture V, Jan.1, 1914).

On such foundations the work was carried forward. Eventually other friends joined in as best they could. The guiding beacon was an unceasing sense of responsibility to lay the groundwork for an astrology which clearly and scientifically recognized man's connection with the stars and yet fully respected the domain of his spiritual freedom and dignity. Thus things gradually shaped themselves.

In 1972, at seventy years of age, Willi was invited by a group of young people to lecture at the International Youth Conference at the Goetheanum in Dornach, where he had given his very first lecture forty-one years earlier at the encouragement of Dr. Vreede.

Though the lecture invitations increased and the breadth and depth of Willi's work grew, it was a great sorrow to him that so few people actively took up the development of astrosophy. He saw the great need to draw from the potentials it contained for humanity to face the oncoming trials at the turn of the century in a positive and constructive way. It was this concern for the future of his fellow human beings that enabled him to overcome his natural reserve and speak out of his convictions. He always said that, for himself, living only in the world of research would have been sufficiently satisfying.

In the following years, publication would be limited to the ongoing "Monthly Letters" to subscribers. A portion of these letters (1972–74) would be published as Willi's final book, *Practical Approach Towards a New Astrosophy*. It is in this work that he brought forth his many years of research, first indicated in *Drama of the Universe,* on a spiritual approach to a heliocentric astrology. This was a revolutionary incision into the world of astrology, which opened the way for a spiritual-scientific understanding of the heliocentric Copernican perspective of the universe. The development of this work was a monumental addition to our understanding of the relation of the human being, and indeed of all of earth evolution, to the heliocentric universe. As he later wrote concerning this:

Another perspective which I was able to work out in great detail over the years was the connection of the human being with the world of the stars from the heliocentric astronomical viewpoint. Some people are still strongly opposed to the heliocentric approach. However, Rudolf Steiner pointed out in the lecture cycle *The Relationship of Earthly Man to the Sun,* Lecture IV, Jan. 11, 1924, that this perspective is correct, although it has come to be a reality through a great mistake or failure in evolution.

In the research which I undertook in this direction, it turned out that the heliocentric approach does not cut out the geocentric completely. Rather it proved to be a kind of complementary relationship.

The study of the very slow movements of the so-called "elements" of the planetary orbits—i.e., nodes and apsides = (perihelion and aphelion)—turned out to be extremely helpful in historic research and also in the relationship of the individual to the world of the stars.

In 1973 Hazel Straker was called back to England to tend to her mother, bringing to a close the twenty-five years of working together with the "family" of colleagues. For a time publishing activities were no longer possible, but distribution of the books by mail was maintained by Veronica Helen Moyer, the fourth member of the little group who emigrated to America with the Suchers. Two years later, Willi's wife died quite suddenly. In spite of this, Willi carried through a lecturing commitment in the East shortly thereafter. He then gradually curtailed his travels and focused on teaching closer to home. Veronica cared for the house and continued helping with the star work. Now, toward the end of his life, he came to accept that his work had not been in vain but that he had managed to lay firm foundations, which would be built on in the future. For the next ten years, Willi's home became a center of activity. He was encouraged as people separately and in study groups came to learn of the work. Countless individuals seeking guidance streamed to his home. Quietly listening to each one, he never addressed the tangled web of personal crisis, but rather lifted one's gaze to the cosmos, gently offering pictures of the great, objective Christ Events to shine like rays of light on the path of destiny. During his workshops at this time, Willi repeatedly referred to Steiner's lecture of October 10, 1919, "Cosmogony, Freedom, and Altruism," in which Steiner outlined the tasks for different parts of the world and pointed to the imperative need for a new cosmogony to arise in America. Willi recognized the importance of this work for Americans to awaken, in a realistic way, to their citizenship in the cosmos. Fittingly, he gave his last lecture, at eighty-two, just two months before his death, to the American Studies class at Rudolf Steiner College in Fair Oaks, California, where he often taught. It was on the Christ Events in relation to the founding of America in 1776.

Willi died peacefully in his sleep on May 21, 1985, receiving visitors until that night. Before his death, together with a small group of friends, he initiated a trust, the Astrosophy Research Center,[1] to care for his publications, personal papers, and library after his death. Space was promised in a nearby house where, as of this publishing, this material is being cared for and is available to anyone wishing to conduct research. Obviously, all personal charts were returned or destroyed, but there is much historical material, medical research, and notes on many other aspects of his far-reaching work.

As one of his students later wrote, "Willi Sucher brought a powerful new impulse toward restoring our knowledge of the stars to a level of mystery wisdom. Most important of all, he opened up the way to a new moral consciousness—one that

---

1. Astrosophy Research Center, P.O. Box 13, Meadow Vista, CA 95722

acknowledges the significance of Christ—in that which concerns the profound relationships prevailing between the cosmos, the Earth, and humankind. Today, scattered here and
there around the world, a small but dedicated group of people
has devoted themselves to cultivating Willi Sucher's work, to
helping astrosophy live as a spiritual impulse in our time."

Late in his life, Willi wrote, "Thus I can finally only say that
I was given by destiny great opportunities of discovering and
working out new creative perspectives of the human being's
connection with the stars, i.e., a new 'astrosophy'. I am most
grateful for these opportunities. However, the great question
for me was always, how can I bring this wisdom to the knowledge of humanity? The answer to this question was never easy,
all during my fifty-two years of working in this field. But there
is hope that this work will be carried into the future and find
more and more possibilities of practical and spiritual application in civilization."

It is hoped that the present publication of this book by the
Anthroposophic Press will serve as a step in fostering Willi's
hope for the future of astrosophy.

Compiled by
Jonathan Hilton

PART ONE

The Changing Countenance of Cosmic Cosmology

The revised content of eight lectures given at
Hawkwood College, Stroud, Gloucestershire, England
August 24–31, 1969.

1.

The Origin of Occult Symbols on the Basis of Cosmic Rhythmology

WITH THE PRESENT LECTURES we intend to enter, as far as this is possible given the limitations of time, the stream of the historic development of cosmology through the ages. Thus we hope to find good foundations for a new cosmology and astrosophy, congenial to the conditions of present and future humanity. Nowadays, when we look up to the stars, they seem to be more or less familiar to us, and generally we do not realize that this was not always so for humanity. There were times when the stars were not visible to human eyes. If we go back to the ancient continent of Atlantis, to about the middle of that long period of evolution, we find that the human being was not able to see the stars because the heavens were shrouded in heavy water vapors, not only fog. These water vapors completely occluded the view of the stars. Then in a certain moment they condensed and thereby caused the tremendous water catastrophes that brought the continent of Atlantis eventually to its downfall. This finally culminated in what is known as the last Ice Age. The water that had condensed covered the Earth and froze into tremendous layers of ice and glaciers of which we still find geological evidence and traces nowadays. There was a moment, and we might have to go back as far as 10,000 to 12,000 years B.C., when these heavy clouds or water vapors dissolved and revealed above them the heavens. Thus did humanity behold the stars for the first time. Of course, humanity was, before that moment, not devoid of strong inner experiences. Being limited in their sense-perception, in that way, threw humans back on themselves, so to speak. Combined with a natural clairvoyance this led to an immensely rich inner experience of the divine world and of the elemental world around them. These were experiences that we can hardly imagine nowadays. The vast storehouse of fairy tales gives us a last glimmer of them. And then, in that moment the stars became visible. The Akkadians, an Atlantean sub-race, were among those to see the stars first. They used this great discovery for navigation and were, so to speak, the first colonizing race. They went out into the world. They navigated on the high seas. And they were able to orient themselves on their journeys with the help of the stars.

We must try to imagine that moment in the history of the human race, that experience of great joy when the stars appeared in the sky for the first time in human evolution. The stars appeared to human beings as a tremendous cosmic story-book of everything they had experienced hitherto in their inner life. They could say, "It's there, what for us was a reality until now only for our inner perception, for our inner dream life, is now written in the heavens." Thus they learned to experience the reality of the divine world, of the divine hierarchies and their creative deeds, as it were, in a mirror made by the stars. They recognized ever more clearly that what until then was like a tremendous world of clairvoyant dreams of a divine world, also an elemental world, was now reflected in the heavens, was, so to speak, written down in a big storybook, a big picture book of cosmic size, comprising all the stars. Thus they lived with and into that experience, and gradually they discovered more and more of that world of reflection. The rising knowledge and awareness of the world of the stars was for them the opportunity to make ever clearer that inner dream-world life that they had had before. In a sense they started to put the backbone of a sidereal mythology into those earlier experiences. Thus humanity started to go along the path of enhanced sense experience. It was particularly the sense of sight that changed the physical constitution of the human being in the course of time.

The impressions of the external world, founded on the use of the senses, gained the upper hand, and gradually humanity moved toward times during which it exposed itself increasingly to this perception of nature on the one hand and to the world of the stars on the other. Thus we can understand that, as humans looked up to the stars, they realized many details. They realized, for instance, what we now call the fixed stars. These are the stars which always remain in the same composition in that great storybook of the heavens, or rather, more or less remain in that position. We know that the fixed stars move too. On the other hand these people also became aware of the world of the moving stars, of the planets. For instance, the Moon is quite a conspicuous example of that world of the wanderers in the heavens. Thus we can understand that gradually humans were able to read that great picturebook.

The first impressions must have been of an almost overwhelming nature. Humans were led to discern constellations in the world of the fixed stars. They could give them names on the basis of what "letters" and "pictures" they represented in the "Great Book of the Cosmos." Likewise did they also name the planets by recognizing their workings. This world became clearer and clearer through sense perception, combined with inner, clairvoyant vision. Thus we can understand that humanity gradually recognized and formulated sidereal mythologies. They connected certain groups of stars with definite experiences, even with definite divinities. Looking at a group of stars, they realized that this is an expression of Zeus, for instance, or an expression of Apollo, and so forth.

However, the more we go back in time, the dimmer and more "nebulous" do these sidereal mythologies appear. For instance, we find that Chinese and Tibetan sidereal mythologies have tremendous volume, comprising gigantic imaginations. But, one could almost say, these imaginations are more connected with a world of elemental beings. Dragons appear in the Chinese zodiac, and horses, and so forth. The Chinese and the Tibetan zodiacs are quite different from ours. That humanity, the Chinese, Tibetans, and Japanese—as a matter of fact the Mongolian race as such—is actually more closely connected with Atlantis than is western humanity. A great part of present humanity has come over from Atlantis, is a kind of leftover from Atlantis. But the Mongolian race has a stronger heritage stemming from Atlantis than the rest. Their sidereal mythologies reflect the time when humanity first, or nearly first, saw the stars in the sky. Of course, since that time even the constellations of the fixed stars we see have changed their appearances. We call them *fixed* stars, meaning that they are fixed in their positions. For instance, we have all seen the Great Bear in the heavens, that is, that constellation with a "body" indicated by four stars forming a near rectangle, and with a tail outlined by three more stars. This group did not always look like that in ancient times, and in some thousands of years hence it will again look quite different. Even the zodiac looked quite different to the Atlanteans. Hence, Mongolians had received a tradition differing from ours in the West. In later times, which are reflected in the sidereal mythology of the Greeks, we meet, more or less, the twelve constellations of the zodiac as we now know them. The Greeks associated them with their gods, other civilizations before them with theirs. This was not done arbitrarily. It stood on the solid ground of inner experience and the corresponding corroboration with external events. This knowledge was evolved in the great Mystery-Temple places of the ancients. Much of this has been described in *The Origin of the Zodiac*, by Rupert Gleadow (London: Jonathan Cape, 1968). He quotes the Roman author Manilius, who wrote about the constellations of the zodiac and their associations with the gods of the Greeks, with Pallas Athene, Apollo, Zeus, and so forth. This zodiac seems to have had its origin more in the western part of Asia.

Another traditional heritage was carried by the people who eventually settled in the Norse countries. They recognized in the constellations, particularly of the zodiac, the reflections and memories of the deeds and fate of their gods. They had a zodiac that speaks of Freya, Vali, Saga, Odin, Skadi, Baldur, and so forth. In this way the whole pantheon of Norse mythology was at one time recognized. Of course, these people certainly did not equate the constellations with the divinities. They perceived the constellations only as an external expression, a kind of great chronicle. They "read" in the heavens the stories that much earlier on they had experienced in a dreamlike clairvoyance.

It may be that the wanderers, the planets, spoke to human beings at first more impressively because they move much faster than the fixed stars. The movements of the fixed stars are so minute that it takes, in some cases, thousands of years to notice any change with the naked eye. With regard to the planets it was different. For instance, the Moon is a perfect example of very quick change. It moves through the heavens in front of the constellations of the zodiac. After some time, that is after about twenty-seven to twenty-eight days, it returns to the same configuration of fixed stars. Furthermore, we notice that it moves through phases. At certain times it rises shortly before sunrise. It appears then as a narrow crescent just above the point at which the Sun will rise. A day or two later it will have disappeared altogether. Then, after a few days we may notice it again, but this time after sunset, as a very narrow crescent above the place where the Sun has set. This means that the Moon is waxing. The crescent becomes broader and broader. Finally, there will come a moment when we see the full, round face of the Moon, rising in the east as the Sun sets in the west. This is Full Moon. After that the disc diminishes again. Finally, there comes a time when it stands again just above the sunrise point as a narrow crescent, as in the beginning of our observations. We notice that this whole cycle takes between twenty-nine and thirty days. This is a phase, or lunation, from New Moon via waxing phase to Full Moon, then through waning Moon again into the next New Moon. This is an expression of a definite rhythm, one that coincides with the rhythm of the seasons. The rhythm of the seasons, of course, goes with the year of 365 days, something that we simply notice when living with nature. People in ancient times, who had no printed calendars, learned to look up to the stars as a kind of living calendar. They noticed that twelve such lunations happened in one year, each Full Moon always in a different place. For instance, if one Full Moon occurred in Taurus, they knew that about thirty days later the next Full Moon would happen in the following constellation of Gemini. According to the records we have, they noticed many more things, for instance, Sun and Moon eclipses. They were even able to pre-calculate their recurrence. Of course, this Moon-phase calendar left a few days over in the year of 365 days. They accounted for this by inserting every third year or so a leap month, in order to bring their calendars up to date. Thus the stars were for these ancient people something very practical, something they needed for their occupations. The farmer had to have a good knowledge of the events in the heavens because this was a calendar, which was needed for sowing, reaping, and so on.

We could go through the movements of the planets and discover similar, though longer, rhythms. Of course, the rhythms of the planets cannot be so easily observed as those of the Moon. Nevertheless, by their patience the ancients did observe the movements and rhythms of the planets in the heavens and their correlations to earthly events in nature and in the human being. And their patience rewarded them. For instance, when they observed the planet that we call Venus they noticed that, seen from the Earth, there was a moment (Fig. 1.1) when Venus was in front of the Sun. On certain occasions they even saw it moving right across the face of the Sun as a tiny dark spot. Furthermore, they knew that this took place in front of a certain constellation of the zodiac. Nine months later they knew that the Sun was at position B. Now they noticed that Venus had disappeared into space behind the Sun, getting smaller and smaller as it moved out behind the Sun. After another nine months the Sun had arrived at C, and Venus was again in front of it. With the Sun at D, Venus had again disappeared as at B. Finally Venus moved again in front of the Sun at E.

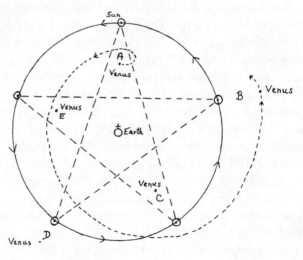

(Figure 1.1)

On the basis of such observations these people must have noticed the existence of a definite regular pattern made by the conjunctions of Venus with the Sun. They must have become aware, by following its conjunctions with the Sun, that Venus draws a pentagram into the cosmos around the Earth. Furthermore, they observed that the cycle we have described so far took four years, whereas in eight years Venus had in each of the five corners (alternately after four years) an inner (inferior) and an outer (superior) conjunction with the Sun. The knowledge of this cosmic fact lies at the foundation of esoteric and occult symbology, of what is known and revered in occultism as the symbol of the pentagram, the five-pointed star.

The ancient sages must have known another rhythm in connection with the planet that we call Mercury (Fig. 1.2). In the course of one year Mercury moves three times behind the Sun. This happens in three different positions of the zodiac, which they seem to have been able to define. Furthermore, the planet steps in the course of one year three times in front of the Sun—in other words, is in inferior conjunction. Thus we can here speak of three cycles in connection with Mercury, each one leading away from an inferior conjunction (in which the planet performs a loop) into a superior conjunction, and back into another inferior conjunction. The three cycles do not actually coincide in time with the year of 365 days. One cycle (for instance, A to B to C) takes only 116 days; therefore the three cycles in one year occur in 348 days. In other words, when the planet is about to return at the end of the third cycle (E to F to G) to the original position, the superior conjunction G (equivalent to A) takes place about seventeen to eighteen days earlier in the year.

Principally, we realize here another cosmic symbol that is well known in esoteric symbology. Mercury draws into the zodiac a hexagon, two triangles intertwined. The one triangle is connected with the inferior conjunctions in front of the Sun, the other with the superior conjunctions behind it. The fact that the corners of the two triangles constantly fall back (caused by the difference between the Mercury year of 348 days and the Sun year of 365 days) is simply an expression of an incessant rotation of the hexagon. To this we must add another fact: the triangle of the inferior conjunctions (B, D, and F) is actually much smaller in the space around the Sun. In these positions the planet is much nearer to the Earth than in the superior events (A, C and E).

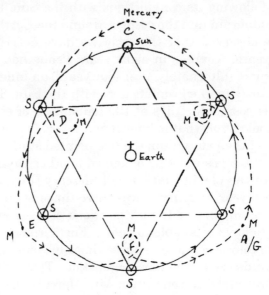

(Figure 1.2)

We can translate this cosmic, geometric figure into a vivid imagination (Fig.1.2b): the bigger triangle can be experienced

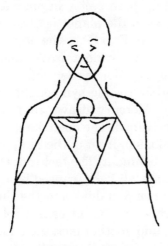

(Figure 1.2b)

as the expression, the external countenance, of a great cosmic being. The smaller triangle can then appear like the imaginative outline of a being who lifts its hands up to the greater one, as it were, in a gesture of supplication. The latter we can depict as a being closer to the Earth, possibly on the Earth, like the human being. The greater one could be perceived as the external expression of a Divine Mother Being in the cosmos who spreads a protective cloak over the petitioner below. This is not too fantastic a picture. Our connection with this planet Mercury is somewhat built on this gesture, in a purely spiritual sense. A very pertinent question is: How do we imagine that the ancient people became aware of these rhythms of Mercury, since the observation of this planet is so very difficult? Yet, on the other hand, the ancients must have known it, as we conclude from their mythologies and knowledge of the working of the planetary genii. We must assume that these people, particularly the sages in the ancient temple-centers, had a power of perception that was essentially different from our present intellect-bound perception. Their still-clairvoyant insight, combined with the scant possibilities of external observation, offered them the possibility not only to be aware of the existence of Mercury but also to know its rhythms and cycles.

There is another rhythm that even in ancient times could easily be observed. It is indicated by the planet Mars (Fig. 1.3).

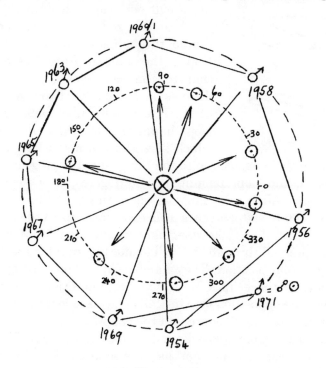

(Figure 1.3)

At certain times it is also standing behind the Sun, as we observe it from the Earth. We speak then of a conjunction of Mars with the Sun. This happens roughly every two years. In between, Mars moves into opposition to the Sun. On such occasions we see the planet at midnight in the zenith above the southern horizon, the south point. At such times Mars is especially bright, shining with a reddish light. This happens also roughly about every two years, a little more than two years. On such occasions Mars performs a loop, moving backwards, or against the direction of its ordinary progress in the zodiac. All the planets normally move in the heavens from right to left as we look toward the Sun, observing them. But when the outer planets—Mars, Jupiter, Saturn—come into opposition to the Sun they turn around and move backward (from left to right), and only after some time do they move straight forward again. In the course of this phase they make a loop, or a hairpin curve (Fig.1.3b). When this takes place again after about two years,

(Figure 1.3b)

Mars will perform its loop further forward in the zodiac. For instance, in March through May 1967, it made a loop in the sidereal Virgo. The opposition to the Sun was at about 205 degrees of the ecliptic. Two years later, in April through July 1969, another loop happened in sidereal Scorpio, which saw Mars in about 250 degrees of the ecliptic in opposition to the Sun. Thus the progress of the opposition, and loop, amounted to 45 degrees, or one eighth of the ecliptic. After this happens eight times in the course of about fifteen to sixteen years, this event returns to the starting point in the zodiac. Likewise, eight conjunctions of Mars with the Sun proceed through the zodiac and return after the eighth to the original position. They happen, approximately, opposite the point in the ecliptic in which the oppositions occur. For instance, the opposition loop in 1967 was preceded in April 1966 by a conjunction at 39 degrees of the ecliptic.

Thus each set of eight events describes an octagon (Fig. 1.3), which one can also present as two squares (Fig. 1.3c), one rotated 45 degrees from the other. This too has become an ancient symbol. It is actually used by astrology as an "aspect," a certain angular relationship between two planets whose relative positions in the zodiac form an angle of 45 degrees. This is the so-called "semi-square." An ancient humanity must have observed all these facts and experienced in these tremendous cosmic rhythms a kind of reflection. For them these must have been pictorial descriptions of what they met as an inner world of awareness of the divine world and also of the world of elemental beings.

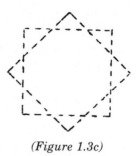

(Figure 1.3c)

Other events in the heavens are associated with the periodic conjunctions and oppositions of Saturn and Jupiter. We have now gone through the whole scale of the planets Venus, Mercury, the Moon, and Mars. They all draw patterns into the heavens, as we saw, that offer the foundations for profound occult symbols. And so do Saturn and Jupiter. Saturn is a very slow-moving planet, needing about thirty years to move once through the circle of the zodiac and return to the initial constellation. Jupiter is much faster, needing only about twelve years to move through the circle of the zodiac. Thus Jupiter needs about one year to move through one constellation. All these facts the ancients could easily observe. On account of the varying speed of the two planets they meet periodically. Naturally, if two runners are running in a circular arena and one is faster then the other, the faster one catches up with the slower one periodically. The ratio of the periodic meetings of Saturn and Jupiter is very interesting. For instance, in the beginning of the twentieth century they met in the course of 1901. (See Fig. 1.4.) That happened in the constellation of Sagittarius. In 1921 they met again in a conjunction in the sidereal constellation of Leo. In 1940–41 another conjunction occurred, this time in the constellation of Aries. One could easily observe it. It happened actually three times, on account of the retrograde movements, the loop-movement, of the two planets. In more recent years, in 1961, another conjunction happened in sidereal Sagittarius. It had returned to approximately the same position as the conjunction in 1901. However, in 1961 it had moved further forward toward the tail of the Archer. In 1981 another conjunction happened, which linked onto the one in 1921. It had moved into sidereal Virgo, whereas the one in 1921 was still on the boundary of Leo and Virgo.

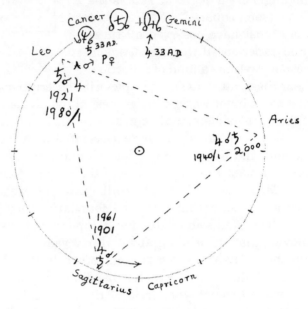

(Figure 1.4)

The meetings of these two planets write into the cosmos an almost equilateral triangle. This was known in ancient times as the golden triangle, and was in a certain sense regarded as a most profound occult symbol. It still appeared in occult publications of the Middle Ages, representing the Divine Trinity, or the trinity of the human being as body, soul and spirit, etc. (See, for instance, "The Secret Symbols of the Rosicrucians" in *A Christian Rosenkreutz Anthology,* compiled and edited by Paul M. Allen, Blauvelt: New York, Rudolf Steiner Publications, 1968.) The Greeks experienced Saturn as an expression of "Omnipotent Father Chronos," Father Time. He took care that the past was never forgotten and saw to it that the threads from the past were woven into the present as destiny. Thus is Saturn an expression of the divine law according to which the universe and humanity were created. He stands as a guardian over the continuation of the divine pattern that was established in the very first beginnings of all evolution. The ancients realized that divine forces spoke through this planet, always calling humanity to attention if there was any attempt to deviate from the divine plan. And they also experienced that these powers brought humanity back into line, sometimes under severe punishment. Jupiter they experienced quite differently. Jupiter, or Jove, was perceived as "Omnipotent Father Ether." Ether is connected with life forces in the cosmos, powers of growth and development toward the future. And the "planet of Jove" would be heard to say, "We must guard that thread of divine law through evolution, but we must also think of the future. We must take the facilities that the present offers in order to develop the future, to lead humanity toward the future. We will not be so severe as Saturn must be. We will give humanity some leeway, be 'jovial.' Perhaps with 'joviality' humanity shall get further." So, Jupiter was realized as a focus of forces in the cosmos that served the human being to realize ideas as much as possible in practical ways. Thus the two, Saturn and Jupiter, combine, and those conjunctions of the two planets have been experienced since ancient times as cosmic conferences. The forces guarding the memory of the dim past and the forces wanting to serve the progress of humanity come together in such a moment. They converse, as it were, about how to go on from a given point in history into the future. And indeed, we can realize that definite events in history are connected with all these dates of conjunctions. Just here in the British Isles and in Europe the years 1940–41 will be very well remembered. These, and all the other dates of similar cosmic events, are connected with moments in history when humanity may be called upon to give an account of its conduct, on the one hand, and on the other hand, to receive new perspectives and guidelines with regard to the future. Thus we can well understand that this triangle of the great conjunctions was often conceived in ancient times as a cosmic sign, or cosmic symbol, of the "All-Seeing Eye of the Divinity," of that which saw to it that the line of evolution was safeguarded and that progress into the future was facilitated.

Thus we discover that many of the traditional symbols that are used in occultism go back ultimately to cosmic realities. This is one of the reasons why we think that a modern, spiritual cosmology is needed, in order to pave the way to a realistic and yet truly spiritual knowledge of the truths contained in esoteric tradition.

2.

The Development of Cosmology in History

LATELY, DEFINITE PROOF has been established of the connection of those great stone circles of the ancient Celtic areas with the stars. The single stones in the circles indicated rising points, for instance, of the Sun at certain seasons and many other astronomical facts that modern humanity has rediscovered only with the help of telescopes and computers. Ancient humanity certainly did not have any of the modern contrivances, yet, it was able to do all this with mental capacities that present humanity has lost completely. This is particularly evident in the ancient monuments in Ireland, at Newgrange, the Hill of Tara, and others. Humanity in the East and Near East also had a close connection with the world of the stars until rather late, chiefly in Mesopotamia. We have definite proof of these correlations by the discoveries and unraveling of the riddles of the so-called ziggurats in the valleys of the Euphrates and Tigris. They were tremendous buildings, resembling pyramids, but terraced, as it were. They carried at the top something like a small sanctuary. This was obviously the observatory of a priest who had been delegated and who was able to follow the courses of the stars. This was not done just for curiosity's sake, but for definite purposes. We have proof of this, for instance, in certain documents of a king who had it written down: "Today I went to the temple of Ishtar (that is, the temple of Venus), and she told me to do this and that," which we would now call matters of foreign affairs of state. Obviously, he had consulted, through his priest, the genii of the stars. Whole communities were thus guided by this knowledge of the movements of the stars, by the rhythms of the planets about which we spoke in the last chapter. The steps of the terraces of these pyramids were painted with colors assigned in those days to the planets. Thus those monuments in Mesopotamia are proof of humanity's connection with the stars in an unegotistical sense. This knowledge was used for the guidance of the nations, right down into the sphere of agriculture, and so on.

In Egypt we discover similar traces. The sanctuaries were built along long axes. There was first the alley of sphinx-like statues. Then one entered a kind of forecourt, after that an inner court, and finally the temple itself. At the far end of the temple was the sanctuary, the chapel that contained the effigies of the deity to which the temple was dedicated. It is an established fact that these long temple axes were oriented toward the rising point of Canopus in the constellation of Argo, which was called the Star of Osiris. The long passage of darkness, the halls, and the courts acted like a telescope. The pyramids had shafts leading from the chambers deep down at the base out into the open. These shafts were not meant to be used as passages because they are oblique. One could not have walked through them. They were also oriented toward the daily orbits of certain stars in the heavens. In these pyramid chambers initiations were obviously performed. For three days the neophyte was put into a deathlike, trance condition, and was even laid into a sarcophagus, deep down under the pyramid, in such a position that upon awakening the neophyte first looked through that long shaft into the external world and saw that star in the heavens. This must have been a tremendous experience.

Then there came a time, and that happened just in Egypt, Mesopotamia, and Chaldea, when awareness of the self awakened. This is described, for instance, in the story of Gilgamesh and his friend Eabani, or Enkidu. The two, so the story tells, had killed the Bull that belonged to Ishtar, the Goddess. These are profound Temple Mysteries. This Bull of Ishtar was really connected with the constellation of Taurus, the Bull. Ishtar was the same as Venus. Venus, as one says sometimes in traditional astrology so glibly, is at home in the sign of the Bull. Why is it just the Bull, and why the connection with Ishtar-Venus? Connected with this were experiences of an earlier humanity, which led to the far distant past of Creation. The Bull which that humanity experienced in the heavens appeared to them as an image of the Creative Word, of the Logos that permeated the world and created the physical object world around us. Now, we hear in this story that Gilgamesh and Enkidu had killed that Bull. What does this mean? They withdrew into their selves, at the cost of their waning awareness of the presence of the divine in nature. Egotism was rising up. It had to rise for the sake of the inner freedom of the human being. Humans began to sever their relationships to the divine intelligences in the world of the stars. Thus the gods eventually "died" in the consciousness of humanity.

In this sense we can understand when we hear Gilgamesh say, "As for you Ishtar, I shall bring you down also. I shall treat you as I have treated the Bull." This is an imaginative description of the moving away from that ancient consciousness that eventually led to the expression we have heard in modern times: "God is dead." He "died" in these peoples' awareness only. After Gilgamesh had spoken those words, which on the surface, sound like terrible blasphemy, Ishtar did not leave it at that. She complained to the highest divinity, and Enkidu, the friend of Gilgamesh, died soon after this incident. Gilgamesh was deeply shattered. He had attained the very first trace of egotistical selfhood. He experienced death as something that meets the individual. An earlier humanity was still well aware that when a human being died only the physical body was put aside, that the soul was all the more alive, was residing in the divine world. To tell human beings in those ages about life after death or about reincarnation would have been futile. They would have replied that they knew that anyway, that it was a reality of inner experience for them. But Gilgamesh experienced, after his friend had died, death as fate, as individual fate. Fate was hitting hard, so to speak, at this man who had taken the first steps of emancipation from the divine world. And so we hear then in the story how Gilgamesh sets out to seek his friend. He is led through many adventures, and we become aware in the descriptions that he goes out into the cosmic world. But he meets there terrible monsters; it is a fearful world. There are scorpions and lions and all kinds of terrible creatures that threaten him. Thus the cosmic world is now already far away from him, a world that is strange, even terrible and fateful. The story of Gilgamesh and Eabani describes the attitude out of which astrology slowly grew as it came down to us. It is that astrology which spoke of the connection of human beings with the world of the stars as a complex that inexorably causes their fate and holds them in pitiless dependence.

In other parts of the world, particularly in the West, the ancient connections with the stars were still carried on right into the first millennium A.D. Even in the second millennium one can still find traces. This is depicted in the story of King Arthur of the Round Table. The childhood of Arthur is described in mighty cosmological terms. (See Fiona Macleod.) The boy Arthur was invited to ascend a huge staircase up to the heavens. He arrived at the seven stars that make up the effigy of the Plough or the Great Bear. There he was met by seven kings. They took him into their midst, and educated and initiated him into the cosmic secrets of the Round Table, which is a vivid imagination of the zodiac. After this experience he descended again and as King Arthur inaugurated the Round Table on the Earth. We hear then the stories of King Arthur and his knights. They were sent out into all the world to help wherever help was required, to protect people who were in danger, and so forth. We see in this a perfect realization of that which was in the heavens as the great and wonderful cosmic order. It was realized on the Earth in quite practical affairs, which concerned the whole human community. King Arthur was also called the Ploughman, because he was deeply associated through his initiation with the mysteries of the Great Bear or Plough. As the king he had to see to it that his people were properly fed. This meant that they had to be educated in the ways of an efficient agriculture, an agriculture that still lived and worked in full harmony with the cosmic rhythms that play into the rhythms of the plant life on the Earth. Therefore it was recognized as a prime necessity to have perfect knowledge of the correlations between cosmos and Earth. The name Arthur seems to have its origin in the Welsh words, *Arth Uthyr,* which simply means "Great Bear." Therefore "Arthur" was probably not a name. It was originally a title, a degree of initiation, in this case into the mystic secrets of the Great Bear or the Cosmic Plough.

This happened in the West. In the East the star wisdom had traveled a different way. It was taken hold of in the end by the egotism of the human being. Individuals wanted to know from the stars their personal fate. They started to ask: How does this or that which happens in the heavens concern me? What will they do to me? Of course, this attitude developed slowly, but in the course of time it became more and more dramatic. All too often one is met head-on in present humanity by this question: What will the stars do to me, what do the star-agencies have in store for me? Unfortunately, this can deteriorate into an attitude of utter fear and helplessness. This was, however, not the end of humanity's relationship to the cosmos. Times moved on and there came the age of Copernicus. Copernicus, who lived during the fifteenth and sixteenth centuries, proposed, but did not really work out in detail, heliocentric astronomy. The Sun, no longer the Earth, was considered to stand in the center of the solar system. Those who came after him worked this out in detail, people like Kepler, Newton, and many others. This step was, in a certain sense, a consequence of that egotism which took hold of the star wisdom in Egypt, etc., which had become astrology. Through Copernican astronomy humanity moved still further away from the stars, until the cosmos was finally regarded as a big mechanism, or computer. With all due reservation, one can say that Copernicanism is a kind of modified Egyptian star wisdom. Of course, in time it radically discarded astrology as sheer folly. But the concept of fate rose to extreme proportions. The Earth was now dethroned. It was considered to be no more than a tiny little planet racing around the mighty Sun.

One can, of course, say: This happened then, and in astronomy. It need not affect us in practical life. Whether the Sun is in the center, or the Earth, this is a matter for the scientists, for the astronomers in their observatories, and it need not concern us practically. However, this is not so. Copernicanism has deeply worked into the social formations and foundations that have developed since. In ancient times we hear of King Arthur's Round Table, of the mighty temple towers in Mesopotamia. There cosmology, or star wisdom, played directly into social life. Whole communities were directed according to that star knowledge. Yet, the same consciousness that formulated Copernicanism also formed the social life of modern humanity. Machiavelli, a contemporary of Copernicus in the beginning of the sixteenth century, wrote the book *Il Principe*. There the monarch is described as someone who should have absolute and unrestricted power. Whatever he does must be correct. He is responsible only to himself and to nobody else. Why should this have any connection with the astronomical world conception that came into being at the same time? The "King" was in ancient times regarded as an earthly representative of the Sun in the heavens. Yet, in all its splendor the Sun was not considered to be the center of the universe, nor was its representative, the King. Thus the very form of the Round Table was also a picture of the denial of prerogatives. Arthur was one of the Knighthood at the Table.

To this, one can object that already long before Christ, oriental despotism had gained great momentum. The truth is that the secret teachings of the ancient temple mysteries had already proclaimed heliocentric cosmologies, whereas in popular conceptions geocentric views were fostered. Evidence exists that this was actually the case, for instance, in Egypt. Because the kings, pharaohs, and so on, were originally initiates of the mysteries, there arose tendencies to build the earthly social orders according to the archetypes of cosmic, heliocentric "Orders," as king-, or pharaoh-, centered communities. And now, simultaneously with the popularization of the heliocentric view by Copernicanism, we witness the fact that Machiavelli creates the philosophical concept of the monarch as the absolute and indisputable center of any social order. And indeed, soon enough, Louis XIV, the French king, entered the historic scene and proclaimed: *"L'etat c'est moi,"* "I am the State," I, the King. The "Sun," in a social sense, had stepped into the center. He and all the monarchs in Europe who could afford it followed his example and built their palaces in a form that was perfectly "heliocentric." Usually these palaces were

built in such a fashion that there
was a central round tower, or a
similar structure with wings on
either side (Fig. 2.1). All the roads
and city streets radiated from this
central point into the environment
like the rays of a Sun. One can
still see many of those palaces in

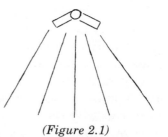

(Figure 2.1)

European capitals. This was "Copernicanism," or the heliocentric system, realized in social formation.

Modern humanity has changed all this considerably. We are
facing now something like an atomic explosion in the place of
the old universe. By the increasing perfection and power of the
telescope we have penetrated into the depths of space. Fantastic distances and worlds beyond our own solar system have
been discovered. Our own solar world has shrunk into insignif-
icance. The greater universe has grown more and more
immense in concept. In fact, it was, and still is, difficult to form
an idea of how big this universe is. We don't know whether it is
finite or infinite. The concept of infinity is painful to a science
built up on quantity. We have discovered solar systems that are
supposed to be much bigger than our own. The central suns of
these systems are, according to the calculations, moving at tremendous speeds. Furthermore, they are moving away from a
kind of center in greater cosmic space with tremendous veloci-
ties, unimaginable by an ordinary human mind. Thus the
greater universe indeed looks like an atomic explosion.

A human mentality that has been led to think in such cosmological terms has had corresponding impacts on modern social
life. If we look at it objectively, we must admit that the forms of
social life, institutions, and so on, of all descriptions are also
constantly facing the danger of atomization. What is the solution? Western humanity is easily inclined to think that complete mechanization and computerization is the answer. Again,
this seems to coincide with the concepts of modern cosmology
that the universe is a vast mechanism. And the suggestion can
easily arise from this that the best and most expedient solution
in human life and in the community is the complete mechanization and computerization of all concerns and institutions.

This is a road that does not lead to solutions but to the sur-
render and abdication of the human race. We cannot possibly
expect that all humanity will accept this, and therein also lies
some hope. Especially the East, that is, western and eastern
Asia, cannot follow this trend, and in this fact are rooted many
of the problems that have arisen between East and West. The
East lives with the dim shadows of a glorious past that speaks,
however faintly, in images of the spirit in the cosmos, in the
human being, in the kingdoms of nature. If we in the West
intend to establish a healthy relationship with the East, we
must not imagine that we can do it by the imposition of com-
mercialism and technology. We can do it only by presenting
ourselves to the East, to global humanity, as a humanity of
spiritual dignity and integrity. And we can attain the means to
do it. If we are looking for an archetype for this we must "go in
quest of the Grail." We will then realize that Parzival, who is
seeking the Grail, is our own image. Parzival is led to the Castle of the Grail. He witnesses all the mysterious events but he
does not ask what they mean. Consequently he is ejected from
the castle. This is the image of the modern human being who
is, just in this age of science, confronted with the riddles of the
universe, of the Earth, of itself. But we do not ask, like Parzival
who wanders for years through the world in deep despair and
hopelessness. Finally, he meets Trevrizent, who brings the
Quest of the Grail back into his consciousness. Now, he enters
the castle a second time and can reach out to the new Mysteries, to answers concerning the spiritual nature of humanity, its
task within the setting of the whole universe, and so on. For
this is the vessel and spirit-content of the Grail. During the six-
ties and seventies of the present century western humanity
passed through the deep dark valley created by its own deeds
in science and technology. But therein also lies the hope that
we may break through, if we can look toward the archetypal
meaning of the figure of Parzival.

As far as we are concerned with our present studies, the
modern consequences of Copernicanism constitute just such a
deep, dark valley. Yet, it would help nobody to go back to the old
good ways of the geocentric world view, for instance, of the
Celts or of Ptolemy. We must go through the valley and find

new portals and roads to spiritual concepts of the universe. Rudolf Steiner pointed out that in the original Rosicrucian schools of the Middle Ages the pupil was first taught the geocentric perspective of the universe. Afterward he was told that this is how it should be, but that in reality it is not so. On account of the great sin of the Earth and its inhabitants—we may think of the Fall in Paradise—the Earth lost its central position. Only after long and strenuous efforts will the Earth in the dim future again be brought into its rightful position. This will be possible only by the acceptance of the Christ impulse. The Christ entered the earthly realm as the representative of the whole solar system. Even as a human being carries in its bodily being an ego that gives meaning and guidance to existence, so is the solar universe permeated by an Ego, and this Ego is the Christ Being. At Golgotha He united with the Earth, so that the Earth may become "Sun" in the far future. In this process all humanity is involved, regardless of race, nation, religion, and so on.

Thus this viewpoint of the Earth in the center is a perspective that must be realized by inner development, by suffering, by rising to the power of the greater, cosmic self. In order to achieve this we must not hesitate to move through the dark valley of Copernicanism. We must break through it to new spiritual vistas. Copernicanism and modern astronomy have given us, after all, a precise mathematical knowledge of the heavens in modern terms. We can calculate the movements of the planets and many other phenomena theoretically thousands of years ahead or back into the past. This is one of the achievements of modern astronomy. It has discovered many details of the movements of the planets that an earlier humanity had comprehended in a kind of clairvoyant, almost dreamlike perception. Previously such knowledge was preserved in the secrets of the mystery temples and was not open to all humanity. In contrast, modern astronomy is a science that is open to anyone who makes the effort to study it. And this very astronomy, although mathematically founded, can give us the means of looking eventually at the solar universe in new, even spiritual, concepts. In ancient times, still in Greece, humanity was able to look up, by clairvoyant perception, to the cosmos and see not only the visible planets but also the spheres of the planets. The Greeks did not experience the planet as a moving entity but as fixed, so to speak, onto a global sphere that was indicated by the orbit of the planet. All these spheres of the planets were conceived of as being concentrically arranged around the Earth. They were rotated by the divine beings who worked in them, and thus also the planets were moved along their orbits. Movement in the cosmos was caused by divine beings, according to this view.

We have completely lost this, because we have lost the concept of the spheres. No modern astronomer cares for spheres anymore. Movement is caused, according to modern views, by purely mechanical factors in the universe. However, it is just with the help of modern astronomy that we can break through to new concepts of the spheres which offer great possibilities. The orbits of the planets are not arranged in perfect circles, nor do they lie exactly on the common, or ecliptical, plane of the solar system. First of all, we have the Sun in the center according to Copernicanism; but it was Kepler who discovered that the planets do not move in circles but in ellipses around the Sun (see Fig. 2.2). This elliptical path brings the planet at certain times into the so-called perihelion (near distance), and at others into the aphelion (that is, the far distance of the planet from the Sun). These elements, as they are called, are the means to assess the inner life of the sphere. As the planet moves along its orbit, it eventually steps into its perihelion, which means that it will be closest to the Sun. At the aphelion it will be far away from the latter. These positions indicate the condition in which the whole sphere lives at a given time. The visible planets are then like "moons" that reflect the life of the spheres. (The spheres are the space-volume inside the orbits.) At the perihelion the planet has a close relationship to the Sun and thereby an affirmative connection with the whole solar system. At the aphelion it expresses, so to speak, the desire to dissociate from

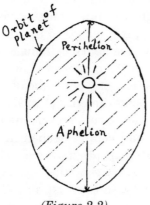

(Figure 2.2)

the latter, to fly away from it into outer space, though it cannot really achieve this because the Sun, even so, holds the family of planets together. Thus we can break through to new concepts of a living universe. Just that which seemed to promote the idea of a purely mechanical cosmos offers a hand to get into new vistas in cosmology.

There are other aspects that have been discovered by modern astronomy. These constitute other elements of the planets. They have been established very precisely with the help of modern computation methods. We take again the Sun in the center (Fig. 2.3). We see the planets move around it, together with the Earth. In the diagram we have inserted the orbit of the Earth, and that of another planet.

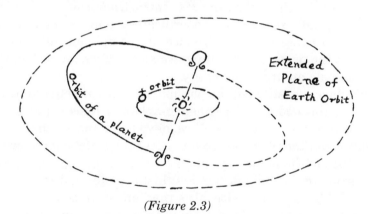

(Figure 2.3)

If we take the planes on which these orbits lie, it may at first look as if they were identical. They seem to move on a common plate, as it were. However, a closer inspection reveals that there are slight angular differences. In other words, the orbits of the planets are all inclined, one against the other. We can now take the plane on which the Earth moves in the course of one year and investigate the inclinations of the other planetary planes in relation to the first. Thus we get crossing-points, or cutting lines between these planes. These are the nodes of the planets with the Earth orbit, or ecliptic plane. Each planet thus establishes an ascending and a descending node. The ascending node is the place—related to the zodiac—where the planet ascends above the ecliptic, the descending node where it descends below. (We have used the symbols that astronomy commonly employs for the nodes.) Of course, these are only mathematical points. We cannot see them. (Neither can we see the perihelions and aphelions of the planets.) We can only calculate them. Nevertheless they are realities of great significance, which refer to the life of the spheres in relationship to the Earth. They indicate that in these places the spheres have the possibility of contact with the Earth and its inhabitants. Thus modern Copernicanism offers us once again the means to come to the concept of a living universe. We can get astronomical tables that give us the precise positions of the nodes, the precise positions of the perihelion-aphelion lines, and so forth. By working and living with these elements—together with certain amplifications that the geocentric view can provide—we can eventually redeem the Copernican system, the heliocentric system, because we can introduce into it an element of life as against the aspects of the universe as a big machine. And this will become an urgent necessity.

We have investigated the possibilities that are offered by the facts of the nodal and perihelion-aphelion lines (lines of the apsides) very thoroughly. And we have found indeed that at the times when the planets step into these lines, the characteristic workings of the planetary spheres and their beings can be well discerned in earthly matters and historic events. However, we should like to emphasize also that on such occasions it is entirely a matter of conscious human awareness and participation whether these events can be employed on Earth constructively, or whether, on account of neglect or rejection, they will work reproachfully and destructively. This is also part, and not the least, of the new comprehension and relationship of humanity to the cosmic world.

3.

Roads to a Modern Cosmology and Astrosophy

IN THE FIRST CHAPTER of this series we spoke about the destiny of astrology as humanity moved toward the modern age. It is a fact that in modern times, about the end of the last century and the beginning of the present, astrology was regarded as utter superstition, as something with which a reasonable human being would not be occupied. However, this is not so anymore. Full evidence exists which proves that science has discovered, or one may say rediscovered, astrology. Some university professors go so far nowadays as to say that perhaps there is something in this astrology after all. Science is on the road to rediscovering astrology on its own ground of statistics. In the following we provide a few examples. As early as the thirties of the present century a paper was published in the United States in which Arthur Fund describes the discovery by Fernando Sanford at Stanford University in California of a correlation between the frequency of visible sunspots and the conjunction rhythms of Venus. It is simply this (Fig. 3.1): We look at it from the heliocentric point of view. At some distance from the central Sun the planet Venus is moving along its orbit. Next to Venus is the orbit of the Earth. Fernando Sanford discovered that during a period of fourteen years, from 1918 to 1932, whenever Venus was—seen from the Earth—behind the Sun (which one calls technically a superior conjunction of Venus), maximums of sunspots were observed. On the average it was a maximum of 714. At the times when Venus had moved into the position between the Sun and the Earth (an inferior conjunction), there occurred a minimum of sunspots. The average was as low as 405. This is an enormous difference. Sunspots appear in a telescope like holes drilled into the surface of the Sun. Of course, the unaided eye sees at best only black spots on the surface of the Sun.

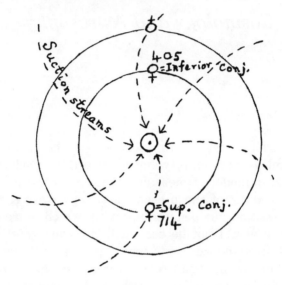

(Figure 3.1)

In these papers, and I shall have to quote a few more in other fields of research, one usually finds one statement, which says: "No explanation can be offered for this or that correlation; it just is so." This we do not find surprising. In order to find explanations in this field, we must take recourse to spiritual science, as founded by Rudolf Steiner, which leads us to the following astronomical conception. The Sun is regarded in astronomy as a body that exerts gravitation on its environment. Thereby it pulls objects toward its surface, and also holds all the planets of the solar system in their orbits. Rudolf Steiner did not agree with this idea. He regarded the Sun, so to speak, as a "hole" in space, a place where positive space comes to an end, reverting into negative space. This may appear a very difficult proposition, but we cannot go into these cosmo-physiological problems here.*

*But will do so later (see pp. 94–95).

Thus the Sun would act like a super pump. The principle of the pump is to create a space that can be emptied of its air content. In this way we can suck, for instance, water into and through such a pump. In a somewhat similar sense the Sun, as "empty" or "minus" space, would draw in from all sides substance from the far distant spaces of the universe toward itself, even from beyond our solar system. This essence or substance is gradually condensed, until it reaches the highest degree of densification on the Earth and on its Moon. Then, as this substance moves on toward the Sun, it is dissolved again. It disappears, as it were, into that emptiness, into that complete reversal of all positive space, which is, according to this idea, the nature of the Sun. So, up to the Earth we would have to expect an increase, beyond the Earth a decrease, of that essence or substance coming from the depths of the universe.

Going along with such concepts, the planets Venus and Mercury, which are closest to the Sun, appear then as stages in the gradual dissolution of space-substance as we move toward the face of the Sun. On the surface of the latter, the final process of dissolution of all positive space into negative space then takes place. Thus we get the phenomenon of the Sun, as we see it in the sky. We have the impression that it is a ball of fire, a place of combustion. The question is where the fuel needed for this combustion process comes from. Science has made strenuous efforts to explain that the Sun is fueled from within. These explanations are not wholly satisfactory. Another explanation would be that the fuel is that "space substance" which is accumulated and finally dissolved on the surface of the Sun. As Venus and Mercury would also be stages on the road of this dissolution, they act like filters that sift the substance coming from the Earth. They filter and prepare it so that when it reaches the surface of the Sun it can more easily be dissolved. Now, we have mentioned above that strange coincidence of a low frequency of sunspots when Venus is in line with the Earth and between the Earth and the Sun. The planet then acts like a sieve and prepares the heavy substance coming from the Earth for final dissolution on the surface of the Sun. If Venus is in superior conjunction, behind the Sun as seen from the Earth, this cannot work. In this case the space-essence, which has gone through the Earth and has reached a high degree of crystalline densification, cannot be absorbed by the Sun. It cannot be completely dissolved by the Sun. We can even think that it hits the Sun, as it were, unprepared, not sufficiently broken down. Therefore it would cause impacts in the outer layers of the Sun and cause the sunspots that then appear as black spots. This would also explain the fact that the sunspots clearly appear in the telescope like holes drilled into the Sun. They would be effected by gross impacts hitting the latter.

Another fact that was discovered is the correlation of radio disturbances, of the so-called radio-weather, the changing clarity of radio reception. For a long time the reasons for this phenomenon were sought. Up to a certain limit an explanation was found in those sunspots of which we have been speaking. But they covered only a low percentage of the radio disturbances. Finally, J.H. Nelson, a researcher for RCA (Radio Corporation of America) one of the top promoters in the field of radio communications, found that the angular positions of the planets of the solar system, seen heliocentrically, are responsible for the radio disturbances (Fig. 3.2). We have here the orbits of the planets. At a certain time we may find two or more planets, say Saturn and Jupiter, at a right-angle position. This invariably causes radio disturbances. There is, of course, the possibility for many such angular positions between the planets. In our present diagram we have an angular relationship of 90 degrees. Naturally, we ask, why should such an angle of 90 degrees influence radio waves on the Earth? In the papers that present the statistical evidence we find again that one sentence: "No explanation can be offered." However, we were bold enough to offer an explanation in another context when we spoke about the rhythms of the planets in Chapter 1. We tried to find the foundations of the ancient occult symbols that have survived into our age. And we discovered that the cosmic facts that underlie the angle of 90 degrees were always well known to astrologers as the square aspect. They also maintained that this is a "bad" aspect. In ancient times this arose out of a real spiritual insight; the modern age has mostly only the tradition left. But why should the "square" aspect be bad?

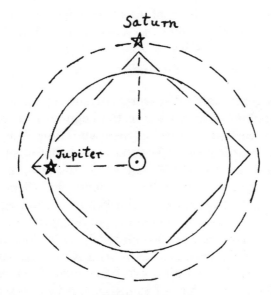

(Figure 3.2)

This can be answered on the basis of the movements of the planet Mars, the aggressor, who has the task of being the aggressor in our solar system. Its conjunctions with the Sun are distributed in the zodiac approximately according to an octagon, or as we earlier described it, according to a double square (Fig. 1.3, on page 35). Therefore "squares" are liable to cause conditions and disturbances as described that fill the whole solar system. We often call such aspects "bad," but this is an unscientific expression. We should develop a more qualified concept, and then we will know what we are handling.

So, this was the discovery of Nelson. It was described in the *RCA Review* of March 1951, and again in an article in the periodical *Electrical Engineering,* in May 1952. After that there occurred an avalanche of discoveries of similar correlations, all verified on the basis of statistics. For instance, one thing that is rather remarkable was the discovery of the interconnection between the phases of the Moon and widespread precipitation. The statistical evidence was collected during the years 1900 to 1924 and then again from 1925 to 1949, from 1,544 weather stations in the U.S.A. They collected 16,057 maximum precipitation records—representing 6,710 individual data. The evidence was published in the form of diagrams, showing the phases of the Moon, from New Moon, to first quarter, Full Moon, last quarter, and so forth. Thus it was established that, indeed, the phases of the Moon are invariably connected with widespread precipitation.

This is a remarkable discovery. Anyone who has lived with these things knows how, even in the beginning of the present century, an earlier science scorned and ridiculed this idea, which was regarded as the superstitious belief of simple shepherds and peasants in the Middle Ages. And now, this very science is obliged to rehabilitate those old shepherds and peasants. On the surface this looks wonderful. But again, there is contained in the corresponding publications that one sentence: "No explanation can be offered." We are left in the dark with regard to the causes. (The official New Zealand Weather Bureau was involved with fifty stations in similar research and obtained the same evidence.) So the avalanche of research went on. A lunar influence on "ice nuclear concentrations" in the upper layers of the Earth's atmosphere was discovered, as well as a lunar effect on the incoming meteor rate. Then it was detected, by similar methods, that the Moon had an influence on geomagnetic disturbances. A good number of papers exist, mostly published in the course of the last decade. Finally, Professor H.S. Burr of Yale discovered (again by statistical evidence) that the lunar phases were connected with the life in living organisms, or rather, and this is a very important point to notice, with the electrical potentials within the living organisms. That is how it is described. Another professor, Dr. Leonard T. Ravitz, also at Yale (all this was published in the magazine *Life*), had measured the "electrical potential" in human beings and detected a fourteen-day rhythm, which often coincided with the lunar phases. Furthermore, New York State statistics at one time seemed to prove that coinciding with the phase of Full Moon there were more admissions into lunatic asylums than otherwise.

We see here also danger signals. The cosmologist might at first rejoice: At last they are coming to reason, they are rediscovering the correlation between the cosmos and the earthly and the human world. But there is the grave danger that all these things are interpreted in terms of material mechanics, electricity, and magnetism. In the end the human being would be regarded as an insignificant and absolutely dependent part of the big mechanism, the big machine and computer, called the universe. And everything, down to the minutest detail of life, would be computed and determined by being injected into this universe.

I should mention another discovery, which was reported in a German periodical. It concerns mongoloid children, a special category of retarded children. A medical doctor in Czechoslovakia, who was originally very skeptical toward the whole idea, eventually decided to cooperate with an astrologer. They worked out the birth charts of fifty mongoloid children and 150 of their brothers and sisters who had the same mothers. First, the geocentric approach gave hardly any clues, but in employing the heliocentric approach they were stunned by the results. They discovered that in all fifty of the charts of the mongoloid children Mercury was in "bad" aspects, according to astrological concepts, for instance, in squares (90 degrees distance) to Venus. At the same time such "bad" aspects showed up between Neptune and the Earth. The brothers' and sisters' charts did not display any such configurations.

In this publication we also search in vain for a satisfactory explanation. Yet, it is very disappointing if one has to work with something that one does not understand, and it can be misleading, too. The question arises: Where can we find guidelines toward a satisfactory understanding? Rudolf Steiner spoke abundantly about the interrelationship between the human being and the cosmos. There is first of all the lecture cycle *Christ and the Spiritual World*, in which he describes, among many other things, the spiritual meaning of the Parzival story according to the Eschenbach version. The latter clearly indicates cosmic correlations. In another context Steiner points out that Wolfram von Eschenbach speaks of Saturn as standing in the constellation of Cancer when Anfortas, the sick Grail King, whom Parzival meets in the Grail Castle, goes through his great suffering. In actual fact Eschenbach says that Saturn was then in the "heights of heaven," which is an indication that it was in Cancer. We have mentioned another transit of Saturn through Cancer elsewhere. (See *Cosmic Christianity,* I.) This happened during the three years of Christ's ministry. There it was also connected with suffering. In the Grail story, Parzival comes and witnesses the suffering of King Anfortas. Instead of asking him for the cause of his illness, he remains silent. Consequently, he is ejected from the Grail castle. Only after long wanderings and long soul-suffering and dejection does he come back, at the moment when Anfortas is going through the culmination of his great suffering. Now Parzival asks him and thereby redeems and heals him. The element of compassion and love, that which entered the world through the Deed of Christ, accomplished the healing of Anfortas. It was at the same time a redemption of Saturn in the constellation of Cancer.

Steiner points out in Lecture 5 of *Christ and the Spiritual World* that after long years of spiritual research he came to the conclusion that we face in our age a kind of rebirth of the ancient astrology of the Third—that is, the Egypto-Chaldean—Age of civilization, of course, on entirely new foundations. That Third Age was the time in the history of the human race when astrology was developed, which has reached right into our present age as tradition. Rudolf Steiner maintains that this new astrology cannot go along the ancient paths and approaches any longer. It can grow only on the basis of a Christian cosmology and perspective of humanity. In this sense Parzival is like the prototype of a new humanity who through his own soul-catharsis grows up to live with the stars, in order to bring healing and redemption to a suffering humanity.

In others of his lectures Rudolf Steiner describes the descent of the human soul from the cosmos into incarnation, and how it is born out of the totality of the universe. It brings down the essence of the zodiac as a form body, which forms the matter that a soul is offered on the Earth through the parents, and compresses it into a human physique. The zodiac represents the great divine image of the human form, which was founded in the most remote past of evolution. (Compare also with Genesis 1:26.) Into this amalgamation of cosmic form body and physical matter the forces of the planets are inserted as foundations for the physiological functions. Thus life, movement, and functions are infused into the human organism. As human beings descend step by step into incarnation, moving through the celestial spheres, they receive from the zodiac the form body, from the planets the ether or life body. And these elements are then finally united with that which the Earth offers. Humans of the present scientific age must learn to comprehend these facts consciously and to realize that they carry the gifts of the whole universe through their incarnations. They have received these gifts in order to transform them, and to increase their "talents." (See also Matthew, 25:14–30.)

Through these correlations with the cosmos humanity is involved in an activity similar to a ritual. As a priest takes the ingredients of bread and wine and offers them up to the divinity for transubstantiation, so must the human being grow up inwardly to a spiritual priesthood with regard to the ingredients that are present in the bodily organization. Rudolf Steiner describes this in the lecture cycle *Man and the World of Stars: The Spiritual Communion of Mankind,* Nov.–Dec. 1922. We, as the present human race, might appear to be still very far away from its realization. But here is a vision of the future when we will be able to handle this. Then, we will take that which is given to us at the moment of incarnation as the "bread" of our physical body and the "wine" of our etheric, or life, body and offer them up for transubstantiation by the conscious evolution of our relationship to the divine. (The physical body, in view of its mineral substance, can be likened to the bread, the etheric body to the wine, on account of its working in the "water," the fluids of the body.)

As we said, the time of the realization of this "cosmic ritual" may be still far away from us in the distant future. But as a kind of very first, inaugural beginning, Steiner gave at the end of the lecture which we mentioned two mantras for meditative work. Eventually, we will need the inner realization of the Christ impulse, because Christ's three-year ministry is the great archetype of this "cosmic ritual." (See *Cosmic Christianity*, beginning on p. 133.) The purely intellectual capacities that we have developed in the course of evolution are insufficient for this work. Through meditative exercises, such as Rudolf Steiner suggested, we will be able to rise to Imaginative cognition, not only to perceive the world abstractly in intellectual terms but to comprehend it by "pictorial-symbolic" perception. (See the basic books of Rudolf Steiner: *Knowledge of the Higher Worlds, Theosophy, Occult Science.*) From there we can rise by certain spiritual exercises to "Inspiration," where we perceive things not only in a kind of "symbolic" realization. There we will be able to "listen," for instance, to what divine beings have to convey to us. At a third stage we can raise ourselves still higher to the capacity of "Intuition," by which we will be enabled to identify ourselves in full consciousness with beings of the invisible world.

Through the development of these capacities, especially "Inspiration" and "Intuition," humans of the future will be able to meet that which is present in their own bodies, essentially, as the essence of the whole zodiac and the forces of all the planets. At the present these are dormant, as it were, in the human organism. But humans of the future will, by their own efforts, develop Inspiration and Intuition and be able to lift the cosmic ingredients out of their dormant state. Thus will humanity, in the far future, transform and restore what is at present lowered, as it were, into the grave of the corruptible body to a realization of that which the Christ reestablished as the resurrection, or cosmic-dynamic, incorruptible body.

Toward the end of his life, Rudolf Steiner described in *Letters to the Members* (of the Anthroposophical Society) in precise terms our new connection with the world of the stars, as he saw it. (English translation, *The Michael Mystery*, Chapter 6.) First he recalls how the world of the stars was experienced by humanity in the dim past differently from today. In the very first beginnings of Creation no stars actually existed. Spiritual-hierarchical beings were, so to speak, in their "places" in the heavens. (See, for instance, *Occult Science*, chapter on "Ancient Saturn.") Then came a time, long ago, when this "World of Being" withdrew and left the first indications of a starry heaven as the revelation of the divine hierarchies. In the end the divinity withdrew so far from the starry heavens that only a memory of the glorious golden past was left in the stars, which even became objects for sense perception. During this modern age, whenever we proceeded to incarnate on this planet there would be no need for us to take recourse to the stars. But now the Archangel Michael, who is called "The Countenance of Christ," has stepped in, insisting that we incarnate timewise so that our individual destinies coincide with the cosmic memories represented in the courses of the stars. And these coincidences are impressed and formed right into our physical and etheric organizations. By his insistence, Michael achieves the reconnection of the dormant but potential spark of the divine presence in the human being, and through this spiritual potential eventually restores the stars to their divine ascendancy. Michael indeed endeavors to raise, through this potential correlation, the redemptive power of the Christ to new cosmic heights. Thus we all have next to us this silent companion, our own relationship to the world of the stars, who participates in all our endeavors, labors, achievements, and possibly also our failures.

At first it is difficult to see how humanity should be able to transform and redeem the stars, even if it is only a matter of the relatively very limited participation of the individual at the moment of incarnation. This perspective changes at once if we take into account the opposite end of humanity's association with the stars, one that is manifest at the moment of death. Thus we can also find an understanding of Rudolf Steiner's additional description in the letter mentioned above. He mentioned that in modern humanity an entirely new relationship to the stars has developed. Through the evolution of potential spiritual culture in ourselves we "feed" that silent star-companion of whom we spoke. Thus do we penetrate the cosmos. In the dim future we will completely transform the present universe, and a new one will come into existence. How can this practically be possible, that the human being, this tiny being, sometimes appearing so miserable, should be able to transform the cosmos? Yet, we can offer some explanation. At the moment of incarnation we receive as human beings the ingredients that we need and of which we have been speaking. From the Earth we receive the mineral substances that are formed into a human body by the cosmic-physical form body, sometimes called the "spirit germ" of the body. Furthermore, we receive from the cosmos, at the last moment before incarnation, the ether body—or life body. In the mineral substances the Earth gives us a "gravity" body, which is bound to the Earth by gravity. But the ether body, which is born out of the totality of cosmic life, infuses levity forces into the physical body whereby life in the human sense is made possible. We are usually not aware of how we constantly overcome the down-pull of gravity. This is the work of the ether body, which prevents us from being a lifeless, static, and heavy rock. It gives us the opportunity of growth and development, eventually of "experience in the sequence of time," or "life."

This experience in time, or life, of a human being is not lost. The whole biography of a human being is, as it were, written down in the ether body. And at the moment of death we give back what we had, so to speak, on loan. The mineral physical substances we give back to the Earth. Likewise, the ether body goes back to its origin in the cosmos, that is, the realm of the planets. Now, it is filled and permeated by our life experiences, by our living biography. During the first three days after death, when the physical body is laid aside and we are still united with our ether body, we experience a great tableau of our whole life. What in Earth-life was spun out in time is now integrated in one coherent picture. This is the ether body, which in this sense one can regard as a time body. After three days this vanishes, because the ether body returns to its origin. It separates from the soul. Still, in it is contained all that we have lived through on the Earth. Thus it goes back to the cosmos and permeates the planets. It does not vanish but lives on in the planets, possibly getting ready to inspire other souls who descend into incarnation. In the course of these deliberations we shall take an opportunity to demonstrate how certain individualities, whom we know historically and also with regard to their spiritual endeavors, may have permeated the planetary world with their ether bodies, or etheric biographies. They were taken up by definite planets, rested there, and were preserved. Eventually other souls, perhaps centuries later, descended, who were possibly inspired by the earlier etherized motives. They took them up, incorporated them into their own ether bodies, and evolved them further in their incarnations.

Thus we can indeed say that we are transforming the stars. Through our ether body we infuse an element into the domains of the planets that was not there before, which matured in our spiritual-moral endeavors. Thereby the planetary world and the cosmos in which we live are gradually transformed. At the end, in the dim future, a new cosmos will have come into existence through the potential evolution of the human race. At present this happens at the moment of death. There will come other times, when we will be able to do this during our sojourn on this planet. Then we will be able to speak to the stars, to respond with spiritual creativeness to their courses. The power to do this we will gain from our soul-union with the Christ Impulse. This will come when we have risen to the reality of the St. Paul experience: "Not I, but Christ in me." Thus can the Second Creation be born. The cosmos around us of the First Creation is moving toward its own exhaustion. It is waiting for that renewal which must come through us, through our new relationship to the stars, which we have tried to outline.

4.

The Archangel Michael and Astrosophy

ON A GREAT NUMBER of occasions, Rudolf Steiner gave practical indications concerning modern humanity's connection with the world of the stars. Particularly in a letter that he wrote on 25 Oct. 1924 (English translation, *The Michael Mystery*), he spoke about humanity's relationship to the stars from the perspective of the mission and the working of the Archangel Michael. He developed there two aspects, one referring to incarnation. At the moment of the approach to incarnation, Michael insists that we orient our descent to the Earth so that our destiny corresponds to the movements and to the rhythms of the stars in the heavens. Practically speaking, there would be, at the present stage of evolution, no longer any need for us to establish this relationship to the stars at our incarnation. That it still exists is due to the insistence of Michael. His intention is to still connect the stars, through the spark of the spirit in the human being, with the evolution of the whole world. Steiner goes on to say in that letter that this achievement gave Michael "such deep satisfaction, that a great part of his life-element, his life-energy, his radiant, sunlike life-will, lives in this satisfaction." But this is not all. Steiner describes a new potential relationship of humanity to the stars. Not only do we take along this star-companion who joined us at birth, and for whom we are expected to make ourselves responsible. As we move toward the future we are called upon to transform that world of the stars through our spirit deeds on the Earth. In the far distant future, we will create a new cosmos. This sounds at first impossibly ambitious. Therefore, I should like to substantiate this in practical detail.

I shall take as an example the birth configuration and the complex of the prenatal star events of Beethoven. Traditional astrology has taken into account for hundreds, perhaps for thousands, of years only the configuration of birth. There exist, however, indications in Egyptian documents suggesting that the Egyptians considered the whole prenatal time, on the basis of the so-called *Trutina Hermetis*. It took into account the approximately nine months before birth, with individual differentiations. The Egyptians claimed that they received this wisdom from Hermes, the founder of their civilization. Finally, the oral tradition was written down by two priest-kings, Nechepso and Petosiris. The documents referring to it are preserved in a library in Paris. With this rule we are led back into prenatal times, close to the moment of conception. It calculates the so-called "Epoch" on the basis of a relationship of Sun, Moon, and Earth. However, the Epoch is not identical with conception.

What then does the Epoch purport? We must look at it as a spiritual-cosmic event. Rudolf Steiner has described it in minute detail. When a human soul comes close to the moment of incarnation, around the time of conception, it is still in the sphere of the Moon. It is clothed in its astral body. Furthermore, it also carries with it the so-called "spirit germ" of the physical body. This is a dynamic extract from the twelve constellations of the zodiac that represents a memory of the divine form body of the human being that was created by the hierarchical world in the dim past. During the major part of the life between two incarnations the soul is engaged, with the help of the divine hierarchies, in reattaining and reestablishing this spirit germ. Then, at the moment of conception, the soul must separate from it. This spirit germ drops out of the soul's hands, so to speak, and unites with the physical germ. It organizes the latter, forces the material that is offered into a human form.

Meanwhile, the soul is still in the sphere of the Moon, at the gateway to the Earth, and is clothed, as it were, in the astral body only. The experience of the loss of the spirit germ then urges the soul to contract the individual ether body out of the cosmic ether. Thus at a certain moment, it stands in the sphere of the Moon, robed in the astral body and the ether body, and now it is ready to descend to the Earth and unite with the physical germ. This happens, as a rule, during the third week of the embryonic development, about the eighteenth day. Therefore, we must distinguish from the conception that moment when the soul draws together its own ether body out of the cosmic ether. This is the Epoch, that which the ancient Egyptians recognized when they followed the Hermetic Rule, or *Trutina Hermetis*.

We shall now draw up the configuration of the heavens, at least in part, during the nine months of Beethoven's prenatal development. This we shall regard as an image of the ether body of Beethoven, which was gradually incorporated into his physical body. Beethoven was born, as far as we know, during the night from the 15th to the 16th of December 1770. In that moment the Sun was just entering the sidereal constellation of Sagittarius (Fig. 4.1). On the basis of our computations of the date of the Epoch, according to the *Trutina Hermetis*, we are led back to the time when the Sun was last in the sidereal constellation of Pisces, about March 22, 1770. During the following nine months the Sun moved through the zodiac, through the ecliptic in a near three-quarters circle. Apart from this, other things happened in the cosmos. For instance, the planet Saturn was moving through the sidereal constellation of Cancer. At the time of the Epoch, Mars was opposite, in the constellation of Capricorn. It was then actually exactly opposite the place where Saturn was at the moment of birth. (All this is taken on the basis of the geocentric view.)

We said that this prenatal chart was a picture of the ether body that united with the physical body. The ether body of the human being keeps the physical body "alive" for a lifetime. It combats the natural inclination of the physical-material substances to decompose, which is nothing else but the natural reaction of material substances that are dissolved, or swim, in water, as they do in a human body. Thus, the ether body has the task of keeping that human body integrated, of keeping it "healthy," as we say. Therefore it must, in a sense, fall in, identify itself, with the spatial form of the physical organism. This we depict in the diagram by the path of the Sun, the three-quarters circle of the Sun. It is a perfect image of the embryo.

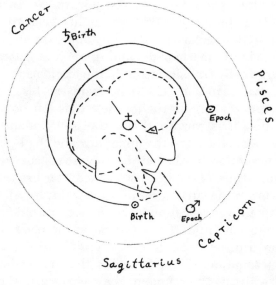

(Figure 4.1)

Corresponding research work over decades has proved indeed that the embryo is depicted in its typical inverted form by the movement of the Sun.

We have drawn, in the diagram, the embryo out of the image of a human head in the background. Why did we do this? We want to infer here that, according to research, the embryo is in the beginning stages predominantly a "head." The trunk and limbs are only "seed points" that develop and expand in later embryonic stages. In this sense the embryo appears as if it was born out of a head. Why should this be so? Why are the typical proportions of the human form from the beginning not indicated? We see deep spiritual reasons for this, associated with the facts that a human being encounters between two incarnations. When a human being dies, the spirit-essence of the physical body, which had formed matter into a typical human form, does not die. It separates from the physical-material body, remains "intact," and is gradually transformed into the head of the following incarnation. How can we understand this? This we try to demonstrate in Figure 4.2. We imagine a human being in the moment of entering the spiritual cosmic world after death and, so to speak, expanding into the widths of space. Certainly, one would not look toward the Earth, and be "inverted" like the embryo. With a mighty gesture one would move out into the cosmos. We can only approximate this in Figure 4.2. This human form of dynamic force that is totally extraverted is gradually transformed into the foundation of the head for the next incarnation. The head from the last incarnation cannot be taken along on the journey. It is left behind on the Earth because it is worn out. But the limbs of the last body become the jaws, the arms the upper jaw and the legs the lower jaw. The foundations of the new brain with its curvatures stem from the intestines of the last body. The latter are transformed and reappear in the curvatures of the new brain. All this is not carried over in a physical-material sense, but only as spiritually dynamic potential. Thus is this spirit body of the last incarnation transformed into the potential form of the head of the next incarnation. It stands behind the embryo form. Out of this head, which is the remnant of a past incarnation, eventually grows the embryo and the new human form. That which had in between been transformed into the archetypes of the jaws now forms the new limbs, and the archetypes of the brain work into the trunk and the inner organs. However, this form is "inverted"; it is looking down to the Earth, to that which is as the umbilical cord the representative of Mother Earth.

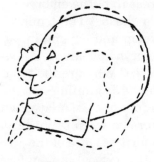

(Figure 4.2)

Compare the cosmic facts that are combined with the archetypal head (Fig.4.1) of Beethoven with a painting of him. The features, particularly the mouth, do not look like those belonging to a happy man. Rather he appears like someone who has a deep grudge. Yet there is also strong determination. We ask, what is the background of all this? We said that we see in the cosmic, prenatal configuration the image of the ether body. The latter carries the karmic memories of past incarnations, though they do not normally enter into full human consciousness. And there must we seek the roots of all that is expressed in the features of Beethoven's countenance. Saturn was in Cancer, an old acquaintance from our series on *Cosmic Christianity*. Saturn was in Cancer at the time of the ministry of Christ. Thus, it is deeply connected with the mysteries of the evolution of the human race. During the three years of Christ's working, He brought redemption to an Earth and a humanity that had fallen deeply. This deep "Fall" is expressed in the constellation of Cancer. Norse mythology speaks of it as the "Twilight of the Gods," which means the twilight of the cognition of the gods by human consciousness. In ancient times the land of the gods, Asgard, was connected with the land of the humans, Midgard, by the Bifrost Bridge. This picture wants to tell us that ancient humankind had, however dim and dreamlike, an awareness of the divine-spiritual world. But the development of human egotism destroyed this awareness. The gods "died" in human consciousness. And thus the Bifrost Bridge that connected Asgard with Midgard was destroyed. This great story is associated in Norse mythology with the constellation of Cancer.

In Greek mythology this is somewhat described in the story of Prometheus, and here is the key we need. Spiritual science indicates that Beethoven was in the dim past incarnated as Prometheus. Prometheus, the son of the Titan Iapetus, decided to assist the human race. He stole the fire of heaven and brought it down to the Earth, to humanity. For that he was severely punished by Zeus. He was chained to a rock and a vulture was set to tear his side open and eat his liver. At night the liver grew again and provided another feast for the vulture. This is but another description of the nature of the constellation of Cancer. Thus Prometheus had to suffer unending pain until much later he was freed by Heracles. It sounds terribly unfair and atrocious that a man who wanted to help the human race by giving it access to fire should be given such a bad deal. To understand it, we must penetrate to the deeper meaning of this picture. The deed of Prometheus was a decisive step on the road toward the emancipation of humanity from the guidance and domination of the divine world. His punishment was the pain of being chained to matter and sickness. The traces of this fate we see in that physiognomy of Beethoven. He carried it all through his life. Yet, there is also that determination which seems to say: I shall go on to bring the fire of enthusiasm and will to the Earth and to humanity. Beethoven did so with his compositions, with the great symphonies, particularly the Ninth Symphony, which speaks even of the spark of freedom in human hearts. But following him also was that destiny of the past. He too was "chained to the rock" of earthly sickness and limitation. Before Saturn had returned, after thirty years, to the place where it was at his birth, he became deaf. This was a most severe destiny for a composer of his stature. Yet, he relentlessly carried on and bequeathed to humanity gifts of unsurpassed beauty and artistic power.

We see this destiny represented in the relationship of Mars and Saturn (Fig. 4.1, on page 69). There is something going on like a tug-of-war between the two, a resurgence of mighty cosmic memories. Beethoven lived with this destiny. We daresay that he himself, already before birth, had decided to take it upon himself in order to give humanity all the more "fire." But the heavenly configuration at the moment of his death suggests that he had conquered it: Saturn and Mars were then (March 26, 1827) in almost the same positions of the zodiac where they stood at the time of the Baptism of Jesus (heliocentric Saturn in the ascending nodal line of Jupiter, and Mars in its own ascending node). We see in this configuration a confirmation that Beethoven had eventually come closer to the real wellsprings of humanity's spiritual freedom, essentially close to the Christ Impulse. Now, we shall look at the other side of our relationship to the cosmos.

For this approach we will study another great personality in modern history, Raphael Santi, the great painter of the Renaissance. Raphael died on Good Friday in 1520, that is, April 6, 1520. At the moment of his death (see Fig. 4.3), the constellation of Scorpio was rising in the east. (The straight line indicates the plane of the horizon in the setting of the zodiac.) Furthermore, on this date Jupiter had just entered Scorpio, and also the waning Moon was not far away, in Scorpio. This is only a part of the whole configuration at that moment, but it is all we need at this point. Scorpio with the poisonous sting is, in a sense, the image of death. Scanty records seem to suggest that in very ancient times humanity clairvoyantly perceived there instead the image of an eagle, a bird that hovers above, that can fly high above the Earth and survey the landscape widely. It was obviously an image of the clairvoyant capacity that ancient human beings had as a natural gift and that transcended the limitations of the temporal. With the loss of those dreamlike, instinctive faculties, the eagle, so to speak, fell deep and became the image of the scorpion with the deadly sting. Yet, this wonderful individuality, Raphael Santi, died under this aspect.

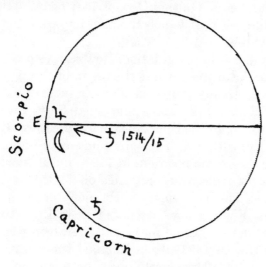

(Figure 4.3)

Why did this happen? It is a splendid exemplification of the fact that humanity's relationship to the cosmos has changed radically. We will be more and more called upon to contribute something new to the cosmos, essentials that did not exist in it before. This can mean redemption and new life even for the cosmos. This sounds impossible, but we can prove it, to a certain extent, with mathematical precision. When a human being dies it means, in practical terms, that life goes out of the physical body. We do not see life as an abstraction. We recognize it as a separate organism, as the life or ether body that separates from the physical body at the moment of death. It works out of a universal memory of evolution, and it also assimilates the whole biography of the human being with whom it was connected. During the first three days after death, this ether body, which is now free from the task of keeping the physical body "alive," is experienced by the soul as a tableau. It contains the details of the whole life. All the events that happened in the sequence of time appear to be collected up in simultaneity. After three days this vanishes from the sight of the soul. In other words, it is dispersed into the cosmos, it is absorbed into the etheric cosmos, which is the planetary world. The cosmos stands waiting for that which comes from the human being. On that Good Friday of 1520, the planets stood waiting for that which Raphael imparted to them as the fruits of his life.

This sounds like wild imagination. How can we prove it? For this we turn to Saturn. Saturn is the organ of cosmic memory, and thus it had already prepared that life tableau during the incarnation of Raphael. When Raphael died, Saturn was below the eastern horizon, in the constellation of Capricorn. By calculation we can know that at a certain moment in Raphael's life Saturn was in this same position, in 1491, and it "memorized" on a grand cosmic scale what happened on Earth, for instance, around Raphael. This is Saturn's "technique" for preparing the etheric tableau. Furthermore, we note that in 1514–15 Saturn was in the positions where Jupiter and the Moon were at the moment of death. In 1491, when Raphael was eight years of age, his mother died. This surely had a profound influence on his development. He no longer had an earthly mother. We can, however, imagine that the soul of the mother helped the growing child from the spiritual spheres of the heavens that she had entered. This decisive stage in Raphael's life Saturn "remembered" while it was in Capricorn. In 1514–15 Raphael went through an important phase of his life. He had been painting many pictures of the Madonna as we know, but in that period he painted the Sistine Madonna, which, in a sense, is the culmination of all his paintings of the Madonna. At the same time Saturn was standing in Scorpio. Among the millionfold events on the Earth in that moment Saturn also looked upon the work of Raphael and collected it up into his great cosmic memory. Then at the moment of death first the Moon and then Jupiter stepped into the places of the zodiac, the cosmic chronicle, in which Saturn had earlier written down his "observations." They received these memories into their being. And now we are suddenly faced with a totally different picture of the Scorpion. We know it as the image of the animal with the deadly sting. And now there appears in the heavens the image of the Divine Sophia. It may sound strange, but this is she whom Raphael depicted in his paintings, particularly in the picture of the Sistine Madonna. She is not just an image of the earthly mother of Jesus. Obviously, she is the Divine Sophia, the Wisdom of the Gods who is carried, as it were, in the crescent of the Moon and reaches out to Jupiter, the planet of cosmic wisdom (Fig. 4.4). The death configuration of Raphael was a perfect signum of the Divine Sophia. Thereby an element of redemption was imprinted into the Scorpion through Jupiter, which may herald a change in the character of the constellation of Scorpio in times to come.

(Figure 4.4)

We can now ask—and this is quite a legitimate question— what is going to happen to this infusion into the sphere of Jupiter and Scorpio? Indeed, it appears to have been well preserved in the cosmos until souls who were prepared could pick up that impulse again and evolve it further. One such soul was the Russian philosopher Soloviev (1853–1900). When he was born, Jupiter was again in the constellation of Scorpio, or rather, during his prenatal development it had moved into that constellation. This we take as a suggestion that Soloviev may have taken up on his return into incarnation the impulse that Raphael had infused into that Jupiter and Moon in Scorpio. Can we find any confirmation of this? Yes, if we study the life of Soloviev, we get a most dramatic affirmative answer. He wrote a poem, "The Three Meetings" which has been translated by George Adams into English. There Soloviev describes his three meetings with the Divine Sophia, or, as he calls her, the Hagia Sophia. As a child he was once standing in front of the picture screen of a Russian cathedral, which separates the altar room from the congregation. There he had for the first time a vision of the Divine Sophia. Later on he studied theology. He also visited England, and on one occasion, when he was sitting in the reading room of the British Museum, he again had a vision of the Divine Sophia. She told him, so he says, to go to Egypt, right into the desert, because there she would reveal herself to him in her full cosmic glory. Soloviev boldly set out for Egypt. People warned him of the dangers but he would not listen. He went out into the desert, a student of theology of those days, clothed in a long black coat and black top hat. Promptly, the Bedouins mistook him for the devil and nearly killed him. However, he survived and lay down during the night in the desert, which was alive with wild animals. Then, in the morning at sunrise he had the most glorious vision and conversation with the Divine Sophia. This experience followed him throughout his life, and through all his studies. His books on religion, on philosophy, on Christianity bear witness to it and would possibly not have been written without that experience. Thus what Raphael had implanted into Scorpio through Jupiter came back. Soloviev was able, by virtue of his destiny and previous incarnations, to take it up and bring it down to the Earth. Certainly, he did not paint the Divine Sophia or the Madonna, like Raphael. But he experienced her in his own ways and elevated her in his soul.

Thus we have here an example that describes how the human race transforms the cosmos, how indeed the constellation of the Scorpion has been permeated with wonderful moral imaginations. And yet, it is only a beginning. More and more will emerge as humanity moves toward the future. Even so, what we have described is not an isolated instance in our present age. There are many more examples that demonstrate the transformation of the zodiac and of the planets. Death, as for instance in the case of Raphael, becomes here spiritual birth. We see it as one of the tasks of a new cosmology to help raise the inner experiences of humanity to heights of precise and scientific cognition of the reality of the spiritual world. So far, we have described only one isolated aspect of the asterogram of death as it is connected with the first three days after death, with the absorption of the ether body into the cosmic ether. There is, however, more. One can recognize in such heavenly configurations the further progress of the soul after passing over, even the weighing of the possibilities of a future incarnation. All this is contained in such an asterogram of "spiritual birth," and with diligent research one can find it. For instance, the time that elapsed between the incarnation of Raphael and of Novalis, the following incarnation of Raphael, was contained, as in a germ, in the latter's star configuration at death. The 252 years, from 1520 to 1772, between the two incarnations are already indicated in the asterogram of 1520, of course, only as possibilities that were offered. The indication is actually of a threefold nature. Moon, Sun, and Saturn present, so to speak, their suggestions with regard to the future, each one in its own way of "timing."

We are not an insignificant coincidence on the planet Earth. A spiritual and really modern cosmology and astrology can indeed reveal that the fruits of human endeavor on this planet can become significant, even creative, in the cosmos. During a long interval of separation from the spiritual cosmos we attained selfhood. With this potential of realization of self, we can again raise ourselves to become citizens of the universe, standing firmly on the Earth and at the same time working with the forces of the universe. Just as we learned to cultivate the soil of this planet, so must we at least begin in our present age to cultivate, plant, and make fertile the fields of the cosmos. These are tremendous perspectives that call for heightened human responsibility and also for the will to work diligently toward the future, even if the present possibilities of realization appear to be so very small. If only a few human beings can under prevailing circumstances accept and live this, then we can have hope and confidence that the road of humanity toward the future will remain open.

5.

Symbology
and Cosmology

TODAY WE SHALL TRY to find an understanding of the symbols that have been used since ancient times for the signs and constellations of the zodiac. They were formed out of a deep knowledge and insight; they have not just been chosen at random. Some of them have certain resemblances to the imaginations that stand behind the constellations. Others seem to be rather obscure. We may even have to modify some of them in a modern sense. This will lead us to realize how these constellations of the zodiac have been transformed by humanity in the course of history. Thus, without fully knowing it, we have contributed toward that spiritual-moral fulfillment of our relationship to the stars in the sense of the suggestions of Rudolf Steiner: Once in ancient times "the stars spoke to man." Then they became silent, and this silence can cause deep sorrow. Yet, amidst this deepening silence "man" is called upon to "speak to the stars." Thereby we can realize our spiritual "humanhood."

(Figure 5.1a)   (Figure 5.1b)

The first symbol is that for Aries (Fig. 5.1a). One can perhaps see in it the horns of the Ram. Yet, this seems to be too superficial. Furthermore, it would not give us very much of a lead with regard to an understanding of the quality of this constellation.

The constellation of Aries is connected with the cosmic arche-type of the human head. It is, in a sense, the root-being, the commencement, of the zodiac. In medieval descriptions and presentations we see the Ram lying on the ground, looking backward toward the crowd that follows him. There are the Bull, the Twins, and so forth. The Ram is thus the leader of the host of the zodiac. Contemplating the association with the human head, we come to ask ourselves: Is this the end? Is there no further development? The human head seems to be fixed and settled forever with regard to its form and function. Is there no higher, or future, purpose concealed in that human head? So far, its present purpose is to serve as a region, an area in the human body in which the main sense organs, not all the sense organs, are concentrated: sight, hearing, taste, and smelling. Also located there is the brain, the organ of intellec-tual thinking, so to speak, of that statistical thinking that rests on the sense perceptions.

We may now ask: What is the meaning of all this? How did this come into existence? In order to discern the direction in which we may find an answer, we look at the plant. The plant can be regarded as a "brother" of the human being that has been left behind by the human race somewhere on the road to humanity. Therefore, it can possibly tell us something about the evolution of the human form. The plant's organization is diametrically different from the organization of the human body. It stretches its roots into the soil. This root-organism can be conceived as the equivalent of the human head. Its limbs—that is, the stems with the leaves and eventually the flowers—are oriented toward the heavens. They receive the heavens into the being of the plant. Thereby the seeds, the means of propa-gation, are finally matured. The human being stands on the Earth, with limbs, and the whole metabolic system, pointing toward the latter. The head is oriented toward the heavens, although it is completely closed off by the skullcap. Yet, there also exist suggestions of the human being's being rooted in the cosmos. For instance, the Moses of Michelangelo shows him with two "horns," or rays, coming out of his forehead. In this way Michelangelo wanted to indicate that, according to the old wisdom, Moses had developed more of a connection than usual with the cosmos. He had developed a certain organ for spiritual perception. In the eastern conception it is called the two-petalled lotus and is centered near the pituitary gland. As much as this refers to a past stage of history, we can see in it also the future of the human head. Moses inaugurated the stream in humanity that had to come right down eventually into intellectualism, into the possibility of objective perception of the world. It was a necessary preparation for objectively seeing the events in Palestine during the three years of Christ's ministry. Apart from this there is also indicated in this Aries "organ" the potential of a new connection with the universe, of the development of organs of perception beyond the physical sense organs. These lotus organs, or chakras according to the eastern conception, can be evolved by every human being who makes the corresponding efforts in the field of meditative and contemplative discipline. Rudolf Steiner has given ample and sound advice for this. Through such inner development we can find again a real connection with the Spirit Cosmos, and evolve our head, or "Aries" organism, to become a root-being in a totally new sense.

Naturally, the realization of this is one of the most difficult things facing humanity. Single personalities in history like Emerson or Grünewald, the painter of the Middle Ages, have moved in this direction. Particularly the latter's painting of the *Isenheim Altarpiece*, which shows in sequence the Birth, the Crucifixion, and also the Resurrection, is a representation of that which Aries can become through humanity in the future. When Grünewald (1528) and Emerson (1882) died, Saturn was in the constellation of Aries. Contrary to birth, we see in the positions of the planets at death a picture of the results of a human life, of achievements, and possibly also of defeats. Saturn would depict more that which was in line with the concerns of humanity, which is assimilated into historic conscience for future perfection or redemption. Another two remarkable personalities who died when Saturn had entered the constellation of Aries were Dostoevski (1881) and Shelley (1822). Thus we can say that this constellation, as the archetype of the human head, is an image of the First Creation (in the sense of Genesis). This aspect is somewhat portrayed by the traditional symbol that we use and that indicates an influx from above into the "below" (Fig. 5.1.a). But in modern times and particularly for the future it should be transformed. This however is entirely given into the hands of the Moral Imagination and Intuitive Creation of the individual human being. In order to meet this aspect of Aries, its symbol would have to be reversed (Fig. 5.1b), so that it indicates a blossoming out, an active opening up toward spiritual heights, toward that which is the greater cosmic picture of our being.

(Figure 5.2a)   (Figure 5.2b)

For Taurus, the Bull, we use the symbol in Figure 5.2a, which comes to us from ancient days. We see in the horns on top of the circle something like cornucopias, containing the totality of the cosmos which has been borne into the lower world, the world of matter. Indeed, in the ancient sense, Taurus is deeply connected with the creation of the material-physical world out of the power and the being of the Logos in the universe. Therefore Taurus is also associated with speech. Originally it was the Cosmic Word that created the world. Here too we can ask what a possible transformation could be. This must have something to do with an understanding of the spiritual origin of this world of matter. Furthermore, we would expect here the eventual development of creative means, in the sense of word and sound. We can indeed find evidence for this. For instance, Konrad Ferdinand Meyer, the great Swiss novelist and poet, or "Master of the Word," died (1896) when Mars, Earth, Neptune, and Pluto were in Taurus (heliocentrically). At the time of Tennyson's death (1892), Venus, Neptune, and Pluto were in Taurus (sidereal), the latter two close together. The German philosopher Schelling died (1854) when Saturn and Venus were in Taurus. Schelling was first concerned with a "Philosophy of Nature." From there he proceeded to write and to present to his contemporaries a "Philosophy of Revelation." Thus he turned from a philosophic contemplation of nature to an understanding of the working of the spirit in nature and in the human being. Furthermore,

Thomas à Kempis died (1471) when Saturn was in Taurus. He was probably the author of *The Imitation of Christ,* a famous medieval mystical book, which has been translated into over fifty languages in more than six thousand editions.

We find Uranus in Taurus in the death asterograms of two great personalities. One is Gotthilf Heinrich Schubert, a German philosopher (d. 1860), natural scientist, and in a modern sense, a psychologist. Apart from all this he was also an eminent cosmologist who wrote books about the interrelationship between microcosm and macrocosm in which he tried to describe the spiritual secrets of Creation from out of the cosmos in a truly Taurian sense. Thomas Aquinas, the great teacher of Scholasticism in the Middle Ages during the thirteenth century, died (1274) when Uranus was in the constellation of Taurus. These people were firmly standing in the world of material reality, but they tried to find the spiritual Logos-aspects behind the external material existence. On this basis we would slightly alter the symbol for Taurus (Fig. 5.2b): a tremendous upper or cosmic world permeating the lower or physical world.

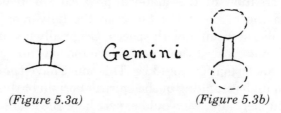

(Figure 5.3a)   (Figure 5.3b)

The constellation of the Twins, or Gemini, reminds us of hierarchy in any sense, the great hierarchy in nature, heaven, and earth, hierarchy in a spiritual sense, the divine hierarchies. We would also see in it hierarchy in a social sense, particularly social foundations and social orders in ancient times, entirely built up hierarchically, for instance, in Asia before the time of Christ and even surviving into later times. It is a principle that has and still is working as an obstruction to progress and is responsible for much of the present disturbances in Asia. The Twins must be transformed. The vertical element—above and below—that is expressed in the ancient symbol of the Twins (Fig. 5.3a) must be combined with the horizontal, that is, the right and left, the side by side. Inasmuch as human community is concerned, this means the establishment of brotherhood. This can be expressed by substituting for the ancient symbol of Gemini one that is tentatively like a cross (Fig. 5.3b). This would indicate "two" but in such a fashion that the vertical is harmoniously combined with the horizontal. Among personalities who entered the spiritual world when Saturn was in Gemini was Beethoven (d. 1827). From Beethoven's life work we can get the impression that he was standing in a kind of "brotherhood" relationship to the spiritual world. Out of this "nearness in a horizontal sense" he was able to create those mighty compositions. William Blake died when Saturn was in Gemini (1827). He had a remarkable, "brotherly" relationship to the invisible, celestial world. It seems to have been awakened in him after his brother had died. In life his brother had been standing next to him. It was a "side-by-side" relationship, as can be expressed in a horizontal Twins symbol. After his brother had entered the invisible world through death, there was added the vertical, the expression of the one being "above" in the heavens, the other "below" on the Earth. William Blake was able to express this in his paintings and in his poetry. Then there was the strange phenomenon of the American Edgar Cayce, who died when Saturn was in Gemini (1945). He was quite a simple man with no academic education. At a certain moment of his life he discovered that he could go at will into a deep trance. A book that has been published about him calls him the "Sleeping Prophet." In this trance condition he was able to tell things of remarkable content and truth. Many sick people were brought in to his presence, and in that state of deep sleep he was able to suggest diagnoses and also therapies of remarkable accuracy, which are used by quite a number of medical practitioners in America at the present moment. When he woke up, he did not know what he had said in his trance condition. This was a unique and singular case which cannot be repeated at random, but as a phenomenon it falls in line with the possible potentials of the constellation Gemini. Rembrandt died when Jupiter was once again in Gemini (1669). He handled most creatively the problems of light and darkness in his well-known paintings. This is again a "Twins" proposition, to make light and darkness stand side by side harmoniously, to let the one carry the other.

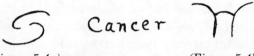

(Figure 5.4a)   *(Figure 5.4b)*

Next we come to the constellation of Cancer. We use the symbol of two spirals, which have come apart (Fig. 5.4a). A rent or a break has taken place between two principles. In a broad sense one could regard the one as standing for involution, the other for evolution. Between them is a deep gap. This describes the old Cancer, that is, the Cancer of which Norse mythology speaks, representing the Bifrost Bridge that once upon a time connected Asgard with Midgard. This myth wants to express the fact that in ancient times the dwellers in Midgard had access to the realms of the spirit, to the gods. This bridge was destroyed in the great battle described as the "Twilight of the Gods," which marks the loss of humanity's insight into the spiritual world. It was this big battle between the gods and the antiforces that was led by Loki, the Evil One. In *Cosmic Christianity* we have described how this great rift was repaired and redeemed by the Deed of Christ. John the Baptist recognized this when he said, "Repent ye, for the Kingdom of Heaven is at hand" (Matthew, 3: 2). The "Sermon on the Mount" reveals these mysteries in great detail (Matthew, 5 and 6)—this "building of new bridges." The new Cancer demands that we transform the old symbol. We can, for instance, imagine two mountain slopes and a deep abyss between them (Fig. 5.4b). Something like a bridge built across would connect the slopes. This is, of course, only a tentative suggestion. We may find in time quite new symbols with regard to this Cancer.   However, even this "building of bridges" has practical perspectives. When Christopher Columbus died (1506), Saturn was between the constellations of Cancer and Leo. He had laid the foundation stones of the bridge from the old to the new continent. He had suffered for years and decades until he reached his goal. This is a kind of living description of what Cancer can be in a very practical sense. When Michelangelo died (1564), not only was Saturn in Cancer but also Jupiter. It was as if the past and future of evolution were in that moment involved in a cosmic conference. Michelangelo took for his artistic expressions the hardest material he could find on Earth, marble, and transformed it into wonderful presentations, especially of the Christ events. Thereby he established bridges across the abyss. In his life work the abyss was the hard material substance that he used, the rocks of the Earth, that which has fallen deepest and furthest away from the spirit. This he transformed into presentations of the great Deed of Golgotha. The German poet Novalis died at a moment when Jupiter was in Cancer (1801). Among his works there are at least three volumes of a kind of diary. He speaks in them about practically everything that a human being can meet on this Earth in the spheres of science, of art, and of religion. With a few words he is able to give the most inspiring insight into these spheres, and he builds thereby bridges of understanding. Paracelsus died when Mars was in Cancer (1541). This great medical doctor of the Middle Ages endeavored all his life to build bridges between that which occurs on the Earth as the phenomenon of human illness and that which works as up-building and creative forces in the cosmos. He did it in order to find effective means of healing. He says, for instance, in one of his books: "That science is truly magic which is able to bring the forces of heaven into a medium and to set them into operation through the same." Or: "It is indeed possible for a man to get hold of and to enclose the whole of the universe in his grasp and this with all its foundations and in clear perception of its perfect entirety" (*Astronomia Magna*). Thus he tried to build bridges between the cosmos and the earthly world.

(*Figure 5.5a*)   (*Figure 5.5b*)

For the constellation of Leo we use the symbol in Figure 5.5a. In ancient times this was the symbol of going out from the Earth into the periphery and eventually into the cosmos. This must also be transformed in our present time and still more in the future. So far, we can only think of a reversal of its dynamics. We may imagine taking the periphery into the center (Fig. 5.5b), even right into the human heart. In order to become fully

human, we may need a deeper than hitherto available understanding of the cosmic forces at work, and also an ability to handle them constructively so that the human being can attain real
"humanhood" on this planet. There is an illuminating example
in history. Saturn was at the moment of Goethe's death (1832)
in the constellation of Leo. His universality and his bringing
this new universality, or "periphery," into earthly life experience
is an expression of the new Leo motif, at least in a germinal
form. Another man who tried to found a knowledge of cosmology, of the workings of the cosmos in earthly substances, was
Culpeper, the famous British herbalist. He died when Saturn
was in Leo (1654). During a relatively short earthly career of
only thirty-eight years, he endeavored to discover the workings
of the planets in plants that were used for medical purposes. He
published heavy volumes of illustrated books in which he
described many medicinal herbs. At the same time he gives in
every case what he considers to be the corresponding relationship with the planetary world. Also that strange personality in
history, Nostradamus, died when Saturn was in Leo (1566). He
had a rather difficult destiny. He had actually trained as a medical doctor and did some good work. However, he became famous
through his prophecies, which he wrote down in verse form,
known as the *Centuries*. They were strangely veiled. Many people have tried to decipher the meaning of these poems. In some
cases one realized their implications only after certain events
had actually happened. The Great Fire of London in 1666 was
one of these instances. After it had taken place someone suddenly noticed that Nostradamus had actually prophesied it during the preceding century. In the same veiled manner he had
prophesied many other things that, when they had been written
down and published, were complete riddles. He received these
prophecies in a strange way. It is said that he lived in a house
that enclosed a central courtyard. There he sat in the deep of
the night after his household had gone to bed and meditated on
the reflections of the stars in the water of the pool that was in
the courtyard. Through the reflection, the light of the stars was
broken, and in the process of their breaking he received those
strange revelations written down as his prophecies. Also Gotthilf Heinrich Schubert, whom we have mentioned above, died

when Saturn was in Leo. We have spoken about him in connection with Uranus in Taurus. Furthermore, Jacob Boehme, the well-known German mystic, died when Saturn was in Leo (1624). He is a remarkable personality in the whole family of the medieval mystics and theosophists. By profession he was a simple cobbler, and yet he wrote the most profound books about spiritual and mystical matters. He well succeeded in uniting the spiritual "periphery" of the human being with the mundane world in which he has to live and which is indicated in the Leo symbol by the little circle.

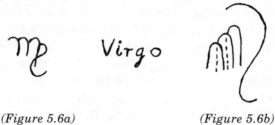

(Figure 5.6a)   (Figure 5.6b)

For Virgo we usually use the symbol in given Figure 5.6a. What does it mean? It seems strange. One can learn to understand it if one studies it together with the symbol for Scorpio (Fig. 5.8a). In between is, of course, Libra. Both symbols, for Virgo and Scorpio, seem to be variations of the letter "m." However, we suggest transforming them into pictures of three successive gates that must be traversed and mastered in order to achieve initiation. Thus, in the case of the suggested symbol for Virgo (Fig. 5.6b), we indicate that eventually cosmic wisdom will be reached, signified by the Serpent of Wisdom and Life. Wisdom and life are two sides of the same cosmic element. Above all we find here Leonardo da Vinci. He died when Jupiter was in the constellation of Virgo (1519). A study of his life and his life's work can convince one that he always, in everything he did, searched for the mysteries of life. We need only to take his most famous painting, *The Last Supper.* To study the gestures of the twelve disciples alone is a revelation. They are a unique presentation of the twelve constellations of the zodiac. Particularly the gestures of their hands reveal that Leonardo must have had a deep understanding of the cosmic mysteries and the wisdom contained in the constellations.

Many more historic personalities could be cited who died when Saturn was in Virgo, not least among them the Greek philosopher Plato (d. 348).

(Figure 5.7a)   L ibra   *(Figure 5.7b)*

For the constellation of Libra we use as a symbol that which somewhat resembles a pair of scales (Fig. 5.7a). We suggest something like a sunrise (Fig. 5.7b) or sunset, when the Sun is at the point of balance between light and darkness. As far as the human being is connected with this constellation, we can indeed find it associated with decision and weighing. For instance, the death configurations of three great astronomers of the Middle Ages show Saturn in Libra. The first one was Copernicus (d. 1543), who introduced the so-called Copernican system. Then we find Tycho Brahe (d. 1601), who could not go along with the Copernican view for spiritual-ethical reasons. Furthermore, Kepler had a similar Saturn when he died (1603). During the last few months of Tycho Brahe's life he worked with Kepler, but after Brahe's death Kepler found himself obliged to further evolve the Copernican system. A tremendous historic battle is indicated in these three Saturn positions in Libra. Copernicus very hesitatingly introduced the heliocentric system. Tycho Brahe did not accept it because he could not agree with the Earth's being dethroned from its central position. Then Kepler followed the suggestions of Copernicus because he saw in them a necessity for the development of modern humanity. In times nearer to us we find that Rudolf Steiner died in 1925 when Saturn had just entered the constellation of Libra. Here we are confronted with a totally new reopening of the gates in a totally new perspective of Libra. This is associated with the mysteries cultivated in the sacred temple places. Steiner prepared the road by offering present-day humanity a means of inner development and of catharsis leading to the reality of this spiritual world. Thus we can also come to understand that this constellation of Libra stands as a balancing element between Virgo and Scorpio. Through the gates of Virgo we enter the mysteries of cosmic wisdom and life.

(Figure 5.8a)   *(Figure 5.8b)*

Scorpio also provides three gates, but in going through them we meet something different from what is in Virgo. In Scorpio we can meet the mysteries of the eternal soul of the human being, provided we have learned to face death as the portal to resurrection (Fig. 5.8b).

Associated with Scorpio we find personalities in history who did travel that road. One who has already been mentioned was Vladimir Soloviev. When he died (1900), Jupiter was in Scorpio. It was also there when he was born (1853). The birth aspect led us to see Soloviev's deep connection with the Divine Sophia, the Hagia Sophia, whom he experienced three times in his life. At death this same Jupiter, now inspired by the life-deeds of Soloviev, would proclaim from the heavens the new aspect of Scorpio: no longer with the deadly sting as the old symbol shows it, but with the breaking through of the human soul to the spiritual realities experienced as the Divine Sophia. Furthermore, Saint Bernard of Clairvaux died (1153) at a moment when Jupiter was in Scorpio. And so did Raphael (1520), the great painter of the Renaissance. We expect that both are so well known that no further comment is necessary.

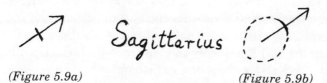

(Figure 5.9a)   *(Figure 5.9b)*

In the symbol of Sagittarius (Fig. 5.9a), a definite dynamism is expressed. It is the arrow that the Archer shoots into cosmic space. We suggest an addition: a little circle from which the arrow arises (Fig. 5.9b). What do we want to indicate by this? The effigy of Sagittarius is shown on old star maps as a centaur. The front and upper part is human. Thus he grows out of a horse's body. We shall now turn this Sagittarius-centaur around, so that the head becomes the Earth and the horse's body moves over the zodiac and the spheres of the planets. But he intends to raise this—possibly still unconscious—embodiment to fully awaken Earth consciousness and creative capacity. Now, we find that Soloviev died at a moment when Saturn was in Sagittarius, and Jupiter was also there at Rudolf Steiner's death. The latter Jupiter in Sagittarius was connected with the birth of Anthroposophy during the years 1900, 1901, and 1902. Available evidence can prove this.

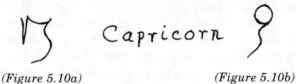

(Figure 5.10a)   *(Figure 5.10b)*

The traditional symbol for Capricorn is somewhat obscure (Fig. 5.10a). In ancient times this constellation was regarded as the portal to the gods. This is clearly expressed by human associations with Capricorn at the moment of their passing over into the spiritual world. Above all Saint Francis died at a moment (1226) when not only Saturn, but also Jupiter and Mars, were in Capricorn (heliocentrically). Saint Francis not only knew but lived a deep connection with the spiritual world. We need only think of his vision of the crucified Seraph who impressed the stigmata upon his body. Christian Morgenstern also died when Jupiter was in Capricorn (1914). Studying his life, which was devoted to Anthroposophy, and also his poetry, can convince one that there was a more than superficial connection with the spiritual world and with a spiritual reality. Swedenborg died (1792) when Jupiter and Mars were in Capricorn (heliocentrically). He does not need any comment in this context. Hahnemann, the founder of homeopathy, died (1843) at a moment when Jupiter and Mercury were in Capricorn (heliocentrically). Hahnemann's own opinion of the nature of that which he had discovered was that by reducing the material substance to a minimum, through the dilution in the homeopathic process, the spiritual forces were freed. Thus, in his opinion, they could work all the stronger and effect cures. This is a demonstration of his realistic connection with the world of the spirit.

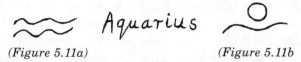

(Figure 5.11a)   *(Figure 5.11b*

For Aquarius we use the symbol in Figure 5.11a, which obviously depicts flowing water. However, this is "water" in the heavens. It speaks of the etheric forces in the heavens. These last constellations, Capricorn and Aquarius-Waterman, and even Sagittarius, are to an extent,. so to speak, reserved for the future. We as human beings are not yet quite so far as to master them. This is because the vernal equinox will reach these constellations only at some future point. Yet, we find single examples that show in which direction evolution may move. When Culpeper, who was mentioned in connection with Saturn in Leo, died, Jupiter was in Aquarius. His association with the world of herbs and their therapeutic properties speaks for his insight into the realm of etheric or life forces. Goethe died (1832) when Jupiter was (heliocentrically) still in Aquarius. We think of his concept, or rather Imagination, of the "archetypal plant," which confirms that he also had a deep insight into this sphere of the life forces. Finally, we mention Thomas Vaughan, the Welsh alchemist, in whose death asterogram (1666) we find Jupiter in Aquarius. His alchemical studies prove that he was a genuine esotericist who had gained access to the formative and creative forces in nature—in other words, to the etheric world.

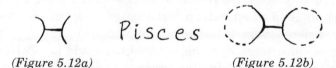

(Figure 5.12a)   *(Figure 5.12b)*

The constellation of Pisces, the Fishes, is close to present humanity because whatever some people may say, the vernal equinox is still below the effigy of the western Fish. Certainly, the waters of Aquarius are already approaching below the present vernal equinox. In other words, we are moving toward a double proposition as far as the latter is concerned. This constellation spans the heavens between Aries and Aquarius. Aries stands for the past, which culminated in the acquisition of head and brain capacities. Aquarius is the uncharted future, which will demand the ability to handle forces of cosmic origin in a healthy way. Between the two stands present humanity, and it is constantly called upon to strike a balance between them, or else face disaster of some kind. This can easily be expressed if we expand the traditional symbol shown in Figure 5.12a to that in 5.12b, showing two worlds united horizontally. Pisces is also the archetypal region of the feet and hands of the human form. They are instruments for achieving independence and emancipation if we compare the human body, for instance, with that of the plant. And it is this independence and spiritual freedom, from all kinds of traditions and dogmas, that modern humanity needs to attain in order to prepare for the demands of the future.

One of the means of achieving spiritual freedom was the development of natural science, although the latter is constantly exposed to the danger of pushing humanity back into un-freedom and bonds that sometimes seem more fearful than those of the past. Thus we can understand that references to Pisces can be found in the *birth configurations* of prominent scientists of the modern age. At the time of birth we see in the heavens the great challenges that come to the human being from the powers active in the spiritual guidance of humanity. Of course, we have thereby no guarantee that the human beings concerned respond to the challenges. This is left to their free decision. Not all whom we shall mention here succeed in this. Kepler, who was mentioned in connection with Copernicus and Tycho Brahe, entered through birth when Jupiter was in Pisces (1571). Isaac Newton was born in 1643 when Saturn and Jupiter were both in Pisces (shortly before their heliocentric conjunction). Faraday, the inventor of the electric dynamo, entered the world in 1791; Saturn was then in Pisces, Jupiter opposite in Virgo. Darwin was born in 1809, when Jupiter was in Pisces. Thus we have here a front-row view of the tremendous battle that is raging in our present age. It is quite obviously a battle of emancipation from old concepts and practices. However, we can succeed only if we are prepared to identify our own existence in freedom of knowledge with cosmic patterns of existence, i.e., with the dynamics of Aquarius. Otherwise our very integrity will be threatened, and we will possibly destroy ourselves. This has become only too obvious during the last few decades.

6.

The Workings of the Planets and their Spheres

IN THIS CHAPTER we shall take a look at the organization and the life of the solar system. Thus we may come to a more thorough understanding of the workings of the planets.

From its commencement, the modern age of science adopted the perspective of the heliocentric, Copernican conception, though Copernicus himself maintained that it was not anything new. He had been inspired to it by teachers who said that similar views had already existed in antiquity. In this sense, for the modern human being, the Sun is in the center of the universe. It is the biggest entity of all, according to this approach, and rules the whole family of the planets. However, we ask, What is the Sun? Obviously it is different from all the other members of the solar family. First, it was taken as a ball of fire, burning away on its surface the substances in its interior, which were either of a solid or of a gaseous nature. The modern age of atomic physics introduced the idea that atomic transformations, what we call on Earth atomic explosions, were taking place on the surface of the Sun. These are supposed to produce the effects that we, on Earth, experience as light and heat.

It sounds, in a sense, simple, but the problems that beset all these ideas are gigantic. Basically, the question of where the fuel for this enormous output of energy is coming from is not really solved. Science has gone all out to suggest many kinds of atomic fission processes and reconstitution of the atom, but these ideas are not really satisfactory. Looking around for some

possible explanation, we come across remarks by Rudolf
Steiner concerning these problems. He confirmed, in a certain
sense, that the processes on the surface of the Sun are a kind of
supercombustion. But in his view this means the absolute termination of all space-substance reality in this universe. And
his fundamental difference from all other approaches is that he
comes to the concept of a Sun whose interior is completely
"empty," i.e., all spatial-physical quality would be eliminated
within the Sun. This is a concept that is difficult to grasp on a
purely physical level, except on the basis of mathematics. It is a
process of complete spiritualization.

In this case the problem of "fuel" is solved. The Sun acts like
a kind of supervacuum, which not only draws in air from the
periphery like a vacuum space in earthly gadgetry, but also
draws in everything that exists in space throughout the
periphery of the solar system around the Sun. This "substance"
is then the "fuel" that is spiritualized in the supercombustion
process.

We shall now try to discern the nature of the planets on the
basis of such a solar system. Figure 6.1 is a partial cross section of the solar system, indicating the orbits of the planets,
with the Sun in the center. This Sun then draws in toward its
"vacuum," by way of that supersuction, substances from the
periphery. They must have undergone tremendous changes as
they passed through the orbits or spheres of the planets on the
way from the periphery. Having first moved through the orbits
of Saturn, Jupiter, and Mars, i.e., through the corresponding
spheres, they arrive eventually in the orbit of the Earth and its
Moon. After that they move on toward the Sun through the
Venus and Mercury spheres and planets. Naturally these
ingredients that are drawn in by the Sun are gradually "condensed" on account of the narrowing of space. Ultimately, on
the Earth, they are compacted into Earth-matter. We indicate
this in the diagram by a kind of rising curve. The rising curve
indicates the increasing density of the substances that are
derived from the "sidereal ingredients," beyond the periphery
of the solar universe. (See chapter 3, pp. 54–55. This is also
described in greater detail in the book *The Drama of the Universe* and is presented in the *Monthly Star Journal.*)

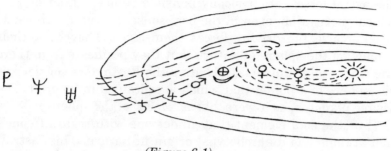

(Figure 6.1)

The idea of rising density as we move from the periphery to the interior of the universe is connected with the gravity on the planets. It is an established scientific-astronomical fact. If we take definite volumes of matter on the Earth, we get an average weight. This weight is, of course, caused by the gravity of the Earth, the pull from the center. On other planets the weight of average substance is less than on the Earth, but varies widely, as astronomy has discovered on the basis of gravitation. It is very low on Saturn, higher already on Jupiter, still higher on Mars. The highest weight is reached on the Earth. Beyond the Earth toward the Sun, it decreases again. So the curve in our diagram "breaks down." The whole appears indeed like a breaking wave. And on the surface of the Sun the ingredients coming, for instance, from the zodiac are then finally dissolved, or rather, spiritualized.

This gives us a splendid foundation to work out the nature of the planets. If we combine this at the same time with Rudolf Steiner's indications concerning the experiences of the soul after death in passing through the spheres of the planets, we can indeed arrive at a perfect cosmography of the latter. Moving through the sphere of Saturn, the soul experiences the archetypes of the astral world, not only astrality itself but also the archetypes of that world. Here on Earth we find, for instance, the manifestations of astrality in human emotions and also in other animated beings. They are then worked out on the purely physical, material level. In the sphere of Saturn, we find the archetypes of all that. For instance, on Earth a great battle, a war, may take place. In that sphere of spiritual-cosmic archetypes, one would see the actual motivating forces, the astral forces, of physically invisible beings behind psychological motivation, even in the best sense. But on Earth we do not perceive the driving forces themselves. They have their "home" in the sphere of Saturn, but they permeate from there the whole solar universe, including the Earth. (We consider as the sphere of a planet that which is contained in its orbit. The visible planet is, as it were, like a "Moon" of the sphere.)

Thus the soul element that guides one with regard to one's motivations has its archetypal origin in Saturn. This "astral" element has even guided the soul into this incarnation. It also has a connection with previous incarnations, with the results of previous incarnations. In this sense Saturn is the archetypal region of soul—or astral—motivation for all physical existence. Therefore Saturn is associated with the skeleton of the human form, with the "motivating" forces of the skeleton, which would be formless without them. Saturn works into the human organism in this fashion, prepares this skeleton "astrally" long before an incarnation, before physical-material realization. Similarly, it prepares also a "skeleton" of the whole incarnation, so to speak, a "karma" or destiny-skeleton, which contains the foundations, the outlines, of that which a human being brings into his incarnation as caused by destiny during a past incarnation.

In the sphere of Jupiter we find after death and long before rebirth the archetypes of the etheric world. Rudolf Steiner describes it in his book *Theosophy* as a "fluid" element that flows as enlivening, archetypal life forces through all living beings who have received their archetypal "motivation" for existence in the sphere of Saturn. Therefore the life—or etheric—forces, which work on Earth, need the actual fluid element as a vehicle. Without water there can be no manifestation of life. This Jupiter element is, in the form of life forces, a motivator of development and evolution. In a cosmically logical sense, that which may originally be only an idea is moved toward the future, toward some form of realization of that idea. Therefore this Jupiter activity in the human being appears as the capacity of thinking, the capacity for realizing ideas.

Then, in the sphere of Mars we enter a domain where we are already drawing close to the Earth. Now, the motivating zodiacal ingredients have been getting denser and denser. Thus, in the sphere of Mars we meet after death the archetypes of the physical object world. We are confronted with the truth concerning the physical objects, also with their illusory nature. When we descend again into incarnation, we acquire in this sphere the capacity to face eventually the world of physical objects that we meet on Earth. Thereby we can maintain our own integrity. The human being is then able to create speech, to develop speech, because speech is that capacity that can face the object world, can describe it by name. It needs a kind of aggressive attitude that is mostly unconscious. We push, so to speak, the objects into their places, withdraw, as it were, into our own integrity. Thereby we are able to use our sense organs and in the end describe the objects.

Before we come to the Earth we enter the sphere of the Moon. There now the final steps toward precipitating the ingredients of the sidereal periphery into material existence take place. It is in this sphere that the soul shortly after death experiences the major part of its purification, or Kama-loka (equivalent of Purgatory). Here all the effects of the final incorporation into Earth matter that occurred at the moment of incarnation must be thrown off because the deepest descent into matter had taken place here. On Earth we see the Moon moving through its phases, through the crescent of the waxing Moon, through Full Moon, and into the crescent of the waning Moon, until it disappears into the New Moon. These are indications of what is happening in the sphere of the Moon. The crescent form suggests a vessel that at certain times receives cosmic substances, and at other times sheds them into cosmic space. Thus at certain times the Moon is turned away from the Sun. It is then exposed to the orbits of the three outer planets, Saturn, Jupiter, and Mars. This is the time when it is waxing and moving toward a Full Moon phase. Then it gathers in the cup of its crescent, as it were, the fullness of those cosmic ingredients that come from the sidereal periphery. They constitute the densified building stones, necessary for physical Earth existence. This is the reason why the phases of the waxing Moon and the Full Moon are observed in agriculture as propitious times for planting, sowing, in fact for everything that needs acceleration of growth, because in that position the Moon can gather up the full supply of building stones and ingredients coming from the zodiacal-sidereal world. Then there come times when the Moon approaches the other side of the Earth, between the Earth and the Sun. It goes then through a waning phase, and eventually through Full Moon. And it has been proved often that such a Moon is unlikely to accelerate the forces of growth. These are times that are far less propitious with regard to planting, sowing, and so forth. Why should this be so? Because we have stepped over to the side of the Earth that is already oriented toward the Sun, where finally the condensed and materialized ingredients of the sidereal world are step-by-step broken down, dissolved, and spiritualized. This periodicity of the Moon works in this sense even right down into the physiology of the human being.

We have now reached the crest of the cosmic wave and the point of its breaking down, from where it ebbs away toward the surface of the Sun. However, the planets Venus and Mercury, and their spheres, are still in between. But they already show less density, or gravitational pull, than the Earth. Thus we have here a process of diminishing substantiality. One can regard them almost as spheres in which the final transmutation on the Sun is being prepared. In this way the human soul, when it moves through the spheres of Mercury and Venus before birth, receives capacities and potentials that enable it in incarnation to refine the life of the soul. Through Venus the human being receives the capacity, of course in degrees and according to karma, to refine its relationship to the environment, both human and beyond the human. Mercury does the same with regard to the human being's intelligent comprehension and actions arising from it. After death the soul is obliged to dissociate itself in these spheres from all-too-strong bonds in these domains of Earth existence.

We may ask ourselves, Why is this process going on? Why is the Sun placed in the heavens, engaging itself, first drawing the ingredients of the sidereal world toward itself, then accumulating and densifying them until the process reaches a climax on Earth? All this, only to break them down again. What is the meaning of it? The meaning of it we can indeed find on the Earth. We as human beings on the Earth are exposed to the physical-material world. It is present in our bodies, which we use as instruments. It reaches a certain perfection though it always seems to leave something to be desired. Then it declines again until we have to give the instrument back to the elements from which it was taken. However, this process, first of growing up and of densification, and then of declining, of dissolution, becomes the foundation for the development of consciousness, even consciousness of self. By meeting the physical-material world, we have a definite sum total of experience which we can only have in connection with our physical material body. But through the impacts that we receive, consciousness, thinking, feeling, and willing are awakened in us. We live in a state of constantly being challenged, and this can become the road to spiritual freedom. We participate in the reception of the densified ingredients coming in on the night-side of the Earth, which is temporarily turned away from the Sun. In other words, we participate in it through our bodies, when we are, as a rule, only in a sleeping or so-called unconscious state. Then, with the rising Sun we face a new day. We wake up and are supposed to be fully conscious. This means that we are gradually involved in a process of dissolving and spiritualizing that which streamed into the Earth at night. Through our labor, both manual and mental, we use up our bodies. They are slowly burned up. Thus, in the daytime, as a rule the opposite side of creation happens, the breaking down of substance, of matter, as it is present in the human body. And thereby consciousness is created; the very first indications of the final elimination and spiritualization on the Sun appear. This process is then continued in the spheres of the planets Venus and Mercury and concluded on the Sun. Therein the meaning of the present stage of evolution reveals itself, which the occultist calls the great cosmic stage of "Earth Evolution." The Earth is, indeed, in a qualitative sense, the center of this universe, as much as we, on modern scientific grounds, insist that the Sun is in the center and the Earth "only" a planet moving around it. On the Earth we can, standing between the impacts of physical-material creation and dissolution, develop ego consciousness. Through the activity of this ego, working through thinking, feeling, and willing, we will eventually be able to lift up external creation to a level where it becomes the power of spiritual creation. In that moment, when this will have been achieved, the Earth will have fulfilled its task. This present universe will dissolve, and evolution will move on to the next stage, to future Jupiter. Future Jupiter will no longer consist of physical-material substance. As much as we can, for instance, detect in the present Earth geological layers, so much will future Jupiter also consist of layers, but they will be layers of thought. The geological layers of the Earth are witnesses to its past. The thought-layers of Jupiter will be witnesses to present-day humanity's thinking.

The great process of creation that we indicated by the picture of the cosmic wave suggests that there may be an inner connection between the planets on either side of the Earth. In this sense Venus would be the solvent of Mars. Mars eventually precipitates the cosmic ingredients into the physical object world. Venus then starts to lift them out of it again. Mars lays the foundation for the world that we perceive with our senses, which we investigate by means of a natural science. Thereby that hard-and-fast physical world is created, in which we have to live. It is split up into uncountable millions of objects. From sometimes harsh experiences we know how often this split-up world causes frustration, antagonism and aggression. In the midst of this stands Venus as an expression and a focus of the cosmic forces of loving comprehension and understanding. They ultimately want to reunite this world that has been split up into numberless objects. The latter are the results of unfathomable ages of past evolution which left the creatures on most varying and therefore unequal stages of existence. The forces of Venus intend to redeem this inequality and to lift up the created objects which were, so to speak, left behind to full and gratifying participation in the process of evolution.

As Venus can be regarded as the solvent of Mars, so Mercury is in a somewhat similar relationship to Jupiter. In the latter we see the representative of cosmic wisdom and cosmic intelligence. In and through Mercury, forces are at work that can help to transform this cosmic intelligence into human intelligence, which then radiates into the hands and feet of the human being.

Finally we come to the Sun, which is in a deep sense connected with Saturn. The latter moves at the periphery of the
solar system, in the sense that it stands at the portal through
which the sidereal substances enter first, in a cosmic psychic-
astral form. Thus in the sphere of Saturn the processes leading
to densification and eventual materialization are initiated. And
on the Sun the final and complete reversal takes place. The
spiritualized essence is then thrown back by the Sun, by its
capacity as a fixed star, into the fixed-star realm, outside our
solar universe. Thus the two, Saturn and Sun, are at opposite
ends of what we may call the "solar process." And yet, considering their meaningful function, they complement one another.

Beyond the orbit of Saturn are those of Uranus, Neptune,
and as far as our present knowledge goes, Pluto. These then
belong to the realms where the sidereal ingredients have not
yet been drawn into the suction stream toward gradual condensation. They are still in a higher, spiritual-archetypal form.
Moreover, these are the spheres that would eventually receive
the respiritualized essences, thrown out again by the Sun, after
the process of dissolution of space-matter has been completed
on the latter. Thus we can understand that those spheres are
not immediately connected with the solar system. They are
something like bystanders and try to speak to the human being
of the invisible worlds, of the spiritual worlds of the divine hierarchies, of all that which is of an absolutely nonmaterial, even
nonspatial, nature. If these messages are not received by the
human being consciously, they can easily wreak "destruction"
and thereby bring home the messages of the temporariness of
all Earth-material existence. Thus they would hope to bring
humanity to a realization of the spiritual origin of all being.

With this approach to the workings of the planetary world,
we shall now take a look at the configuration of the heavens at
the moment of Leonardo da Vinci's incarnation (Fig. 6.2). He
was born on April 15, 1452 (From *Leonardo da Vinci,* Phaidon
Press, London).

To start with, the sphere of Saturn gives a significant impli-
cation. Venus was at the moment of Leonardo's birth almost
precisely in the perihelion of Saturn. The perihelion of a
planet is one of the descriptive astronomical "elements" of the

corresponding sphere. It is the point where the planet, along its orbit, comes closest to the Sun. Opposite is the aphelion, where the planet is farthest away from the Sun. The connecting line between perihelion and aphelion is called the line of the apsides. This line does not stand still but slowly moves, in relationship to the zodiac. Thus the perihelion of Saturn entered the constellation of Gemini around 1400 A.D., coming from Taurus.

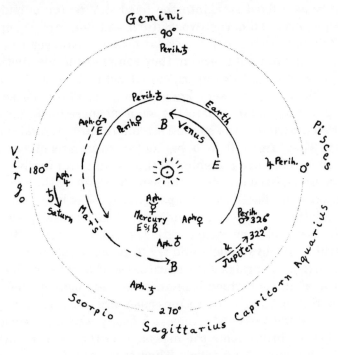

(Figure 6.2)

The line of the apsides is an indication of the "inner mood," as it were, of the planet's association with the solar system. Through the perihelion the corresponding planetary sphere bends, so to speak, sympathetically to the concerns of the solar universe, unites with them. The aphelion of the sphere expresses more the tendency to leave the solar system, to move away and out into sidereal space. The Sun is then always able to persuade the planet not to leave the solar family altogether, by the power that we call, in a purely material sense, gravitational attraction.

Thus we can, with all due reserve, compare the perihelion part of the sphere to the head and sense organs of the human form, which enable the human being to "bend" to the world presenting itself through the senses. The aphelion part could then be seen to be illustrated by the heart-limb-motion organism of the human form.

Saturn is the karmic "mainstay," an expression of the cosmic soul-motives of a human being that are carried into an incarnation. At Leonardo's birth the planet Venus moved into the perihelion of this Saturn, as we said. Venus is connected in the human being with the relationship that is formed to the environment, in the widest sense, with one's particular integration into it. If this combines with the perihelion part of Saturn, it would indicate that Leonardo had built into his organism quite a special potential for a "sympathetic" and feeling relationship to his earthly environment. It would express itself in that kind of duality that is represented in the ingress of the perihelion of Saturn into Gemini since c. 1400 A.D. In Leonardo's life this was present, on the one hand, as the great artist whom we admire. On the other hand, he was the scientist, at the dawn of the modern age, who occupied himself with all kinds of scientific and technological projects. For instance, he tried to build, for the first time, airplanes, though he did not succeed very well.

The sphere of Jupiter was also strongly engaged. At the time of his Epoch, Saturn was close to the aphelion line of Jupiter. (The moment of the "Epoch" is calculated on the basis of the relative position of Earth, Moon, and Sun at birth. It leads back—on the average—to a moment about 273 days before birth, but it is not identical with the physical conception. We receive at that time in the sphere of the Moon our ether, or life, body.) This Jupiter sphere brings the sidereal ingredients one step closer to the final condensation into matter on the Earth. However, it keeps them still on a high level of cosmic-etheric existence. We can grasp them as ideas, through the power of thinking of which we are capable, at least as a start, by means of the nervous system being built into our organism. This is done through the workings of Jupiter. The perihelion of Jupiter has been since long ago and will be for a long time in the constellation of Pisces. It is indeed connected with the long spell in evolution devoted to the development of thinking. The aphelion is in the constellation of Virgo. This is an expression of the need for the activation of thinking, so that it can rise eventually to intuitive thinking, to wisdom (Virgo, the three gates to wisdom—see p. 87).

The association of Saturn with the aphelion of Jupiter in Leonardo's heaven of incarnation indicates that there was a strong potential for "historic" wisdom in him. This came to expression, for instance, in *The Last Supper* where he practically depicted the association of the twelve apostles with the twelve constellations of the zodiac in an unsurpassed fashion. This motif of Jupiter-aphelion-Virgo was further accentuated by a conjunction (heliocentric, of course) of Mars and Venus in that line during the embryonic development of Leonardo. One can say that, thereby, a note of healing was introduced by the all-loving power of Venus into the sense-oriented tendencies of Mars. The artistic creations of Leonardo surely display this healing quality.

The sphere of Mars is connected with the last and decisive stage of condensation of the sidereal ingredients into matter. The elements of this sphere were also significantly engaged in Leonardo's configuration. At birth the planet Jupiter had arrived almost in the perihelion line of Mars, which is in Aquarius. It had entered that constellation in about 33 A.D. Being in this line, Jupiter indicates here the potential for a deep wisdom, regarding the events during Christ's ministry, from 30 through 33 A.D. It seems that it was more than a traditional connection, one based on reincarnation. In this light the *The Last Supper* appears in a significant perspective. Furthermore Mars was around the Epoch (see above) in the aphelion line of its own sphere. In this we would see a reflection of that sense of realism present in Leonardo's life work, in everything he did.

We said earlier that we see in Venus and its sphere the solvent of Mars. Through the impulses of love and compassion it works toward the reunion and healing of that which has been thrown by Mars into isolation and materialization. This impulse was also strong in Leonardo's work and is indicated by the planet Jupiter in the aphelion of Venus, in Capricorn, during the earlier stages of the embryonic development. Jupiter's

wisdom appears here like a mediator between the spheres of
Venus and Mars. (Jupiter was in the perihelion of Mars, see
above.) This was associated with Leonardo's apparent great
love of the Earth and all its properties, signified by the position
of the Earth in the aphelion line of Venus around the time of
the Epoch. One feature of this life of the Venus sphere is very
significant. During the embryonic development of Leonardo a
superior conjunction of Venus with the Sun took place, seen
from the Earth. These conjunctions recur in intervals of eight
years, with intervening conjunctions after four years of the
reverse order, i.e., inferior following superior ones, and vice
versa (see p.33). Thus the superior conjunction of Venus during
Leonardo's incarnation, which was significantly in its own aphelion, is historically connected with the last days before the
Mystery of Golgotha, April 3, 33 A.D. (See *Cosmic Christianity*).
Then Venus was in an inferior conjunction. These inferior conjunctions are connected with the bringing down of cosmic
Venus qualities into Earth existence, as, for instance, in the
raising of Lazarus. The superior conjunctions reflect the gath-
ering-up of the corresponding qualities in cosmic space.

The sphere of Mercury reflects a possible spiritual potentiza-
tion of the qualities of Jupiter, for instance, in human intelligence and activity. We would expect that this sphere also
indicated some remarkable potential in Leonardo's heaven of
incarnation. And indeed, we can discern some outstanding features. Around the time of the birth, Mercury was close to its
own aphelion, in the constellation of Scorpio. The latter is
deeply associated with the descent of humanity into realms of
spirit-removed material consciousness. This is, in a sense, a
death process, and this is what the poisonous sting of the Scorpion indicates. This was necessary for the accomplishment of
the spiritual independence and freedom of the human being.
However, out of freedom the human being must, and will, in the
future find the road that leads back to a realization of the spiritual world and the beings dwelling in it. Otherwise humanity
would lose its integrity and meaning in the cosmic process.
Thus the sphere of Mercury is an expression of tremendous battles in which humanity is involved, battles that concern the
development of intelligence. The question is whether humanity

will be able to lift it up to cosmic intelligence and realization beyond mere existence in matter, or will fall ever deeper into earthbound and short-sighted intellectualism. The latter would cast the human being down into the abyss of nonexistence as a self. The final stages of this battle will occur in the future, signified by the visions of the Apocalyptic Horsemen (Revelation 6), particularly by the red horse. The vernal equinox will by then have arrived in the equivalent of the present constellation of Scorpio.

This great battle lived in Leonardo too, though in a hidden way. We get glimpses of it if we listen to the story of his work on the *Last Supper*. The execution of the painting of Judas, the traitor, caused him great pain and reluctance. This resulted in a long delay in the completion of the painting, so much so that the abbot of the monastery that was to treasure it accused Leonardo of deliberately dragging his feet.

Finally, we have to look at the Sun, which we said is the complement of Saturn. However, in the case of the Sun, as the fixed star, we have no elements of a sphere. In a sense, its sphere is the totality of the solar universe. Therefore it can throw the respiritualized ingredients of sidereal origin right back to the periphery. But we realize one thing if we take a close look at what we called the solar process: The planet Earth is the *qualitative* center of the sphere of the Sun. There the highest degree of densification and the inauguration of the dissolution take place. Thus we can see in the "elements" of the Earth orbit an indication of the life of the Sun sphere. These are the perihelion and aphelion of the Earth, which are at present oriented toward Gemini and Sagittarius. This means that we can see in the line of the apsides of our own planet an expression of potentials toward realizing and lifting up to "I" experience what is inaugurated by the sphere of Saturn as *cosmic*-psychic challenge.

We find in the incarnation configuration of Leonardo that Mars stepped into the elongated aphelion of the Earth at the time of birth. It had started out, at the Epoch, in the neighborhood of its own aphelion.

The aphelion of our planet is, at present, only 10 degrees distant from that of Saturn. They are on the way to moving closer together, but the two lines will not coincide before c. 6000 A.D.

Thus we have here elements that refer us to the last stages of the present epoch of human evolution, the so-called Post-Atlantean Epoch that started c. 7000 B.C., with the Ancient Indian civilization, and which will come to an end in c. 7900 A.D.

The fact that Mars was in the aphelion of the Earth at Leonardo's birth implies a significant potential in his life. On the surface it looks like a defect. In his own opinion he was not able to realize his intentions to full completion. His earthly deeds were attempts in definite directions. But we must look at them from long-range standpoints for the future. Leonardo was, in a sense, preparing the far future stages of human realization. And, although this spells out imperfection with regard to the present, it harbors great hopes for the future. In this lies the greatness of Leonardo, his significance for present humanity, the fulfillment of that promise contained in the combination of Venus with the perihelion of Saturn at his birth.

This is meant only as an example to demonstrate how we can positively and constructively take the spheres of the planets as efficient means of interpreting the relationship between the cosmos and the human being. It should not be regarded as a fully comprehensive delineation.

7.

Turning Points in History—I

THE LAST TWO CHAPTERS will be devoted to an estimation of the situation in which we stand. First, we shall take wider historic perspectives, over the centuries, even millennia. And in the last chapter we shall look at this present age, the present century, and the preparation for the future when the vernal point will enter the constellation of Aquarius.

One rather simple, though not an easy, method is to compare pre-Christian stages of evolution with phases in post-Christian times. For this purpose we will take the Mystery of Golgotha as the turning point. If we go back from 33 A.D. (the year of Golgotha) 2,000 years into the past (Fig. 7.1), or more precisely 1,967 years, i.e., to the approximate time of Abraham, then we arrive at 1934 B.C. Now we go forward into post-Christian times, that is, from 33 A.D., 1,967 years. Thus we come to the year 2000 A.D. as the exact point of reflection of the earlier time, the time of Abraham. What is the historic significance of Abraham? He lived in Mesopotamia, more precisely in Ur. This was probably not the Ur that has been dug up in recent decades. Nevertheless it belonged to that whole cycle of ancient civilizations in the valleys of the Euphrates and Tigris. These civilizations were predominantly built on the ancient star wisdom that

(Figure 7.1)

was then prevalent. In those valleys can still be found the so-called ziggurats, step pyramids of various heights, usually with seven terraces, one built on top of the other (Figure 7.2). As a rule, staircases lead up to the top. On the top terrace was built a small sanctuary, which seems to have been used for astronomical observations. Of course, those astronomical observations, though very precise, were different from ours. The Chaldeans had a very thorough knowledge of the rhythms in the cosmos.

Thus these town civilizations were guided according to a highly cultivated star wisdom. It was brought down more and more into mundane uses. Finally, it was used for personal human purposes. This was the birth of predictive astrology, as we know it in a modern sense, which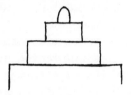

(Figure 7.2)

was more and more used for egotistical purposes. It purported that the human being was dependent on the stars. The stars of birth expressed one's destiny. Thus, by investigating the rhythms, the movements, and the positions of the stars, one hoped to get information about the course of expected events in a human life. Originally, it was the destiny of the whole community that this star wisdom was supposed to discern on this basis. We actually possess documents of that time in which one of the kings of such a community said: "Today I have been to the temple of Ishtar (who is the same as Venus), and she gave me indications and inspirations to do certain things in the interest of the city-community."

These temple towers were attached to the temples, the mystery centers of ancient times. Abraham moved away from such places. He perceived that decadence was setting in with regard to that ancient association of humanity with the world of the stars and that these civilizations were no longer in line with healthy evolution. He experienced it as his task, his divine commission, to prepare a different humanity, different from that which existed in this large valley. That decadence which rose in those localities did not leave any room for evolution toward human freedom, eventually spiritual freedom. Human beings had come to be regarded as being dominated by the stars. To speak of human freedom, even of evolution and the development of spiritual initiative, seems senseless in these contexts. The human being can only execute and demonstrate what is imposed from without. Thus a star wisdom that once belonged to the highest degrees of the mysteries was slowly debased and misused for egotistical purposes and for chaining the human being to "fate." This is the reason why a new star wisdom, congenial to the needs and conditions of modern humanity, must be careful and reserved to the utmost. Therefore we make it so unapproachable and "hang it as high as the stars themselves," so that it cannot be torn down into mundane distortion and degradation.

Abraham moved away from that part of the world (Genesis 12). His task was to build a humanity that was supposed to learn to stand on its own ground. One of the first steps toward this was the development of thinking and the brain-bound intellect. So, Abraham went first to Canaan and from there to Egypt. But in Egypt things did not work out very well (Genesis 12). He returned to what we call Palestine. And there he founded the new line, the new humanity. Neither Egypt nor Chaldea were suitable any longer for the new that had to come. It was possible only in a land between them. We are living at present in a kind of reflection of that age, and one can say that Abraham, of course figuratively speaking, should now "return to Ur," to a new Ur, to a new Christian star wisdom. This can be developed only on the foundation of the modern consciousness of humanity, by a humanity that has realized the perspective and significance, for instance, of *The Philosophy of Spiritual Activity,* or *The Philosophy of Freedom*, which Rudolf Steiner published in 1893. We have the ways and means to move toward a human future that will find again the stars on the foundation of freedom and love. We will again experience, but with full consciousness, the star companions beside us. However, we will realize ourselves to be free, not dominated by the stars. Eventually, in the distant future, we must rise to become a free cooperator with the stars. Rudolf Steiner inaugurated the very first steps toward the attainment of this, especially toward the end of his life. He described then the new creative relationship of humanity to the stars. And this is the foundation on which we must build as we go toward the future.

Thus we can go through history in pre-Christian times and find corresponding post-Christian reflections. For instance, the Gautama Buddha died either in 544 B.C. (version of the Ceylon Buddhists), or 483 B.C. This reflects itself into the post-Christian times of 516—577 A.D., which saw the glory of Celtic Christianity, as it was represented by such a great personality as St. Columba and others.

Apart from these historic reflections, certain cosmic rhythms are important in order to form concepts concerning the interrelationship between human history and the heavens. One such event involves a long rhythm. It is made by the conjunction of *Pluto* and *Uranus*. Pluto was discovered in the thirties of the present century. It needs nearly 250 years to go once through the zodiac. Its orbit is highly elliptical. Thus in one portion of the zodiac it moves very slowly, whereas in the opposite part it moves more quickly. Uranus is almost three times as fast. It needs 84 years and only a few days more, and therefore it sticks almost precisely to the seven-years rhythm, or 12 x 7 years. In other words, it needs seven years to move from one constellation of the zodiac to the following one. Naturally, on account of the slow movement of the two planets and the difference in velocity, they meet rarely. In fact, conjunctions take place in only two points of the zodiac. However, these points where they meet and also the opposition points are rather illuminating with regard to chronology and history. As a rule we study them only in post-Christian times, because the elements of the orbit of Pluto are still a bit vague from an astronomical point of view.

Two points of the zodiac were activated by these planets during the present century (Fig. 7.3). In the beginning, around February 8, 1902, an opposition took place between Pluto and Uranus. Pluto was here in the sidereal constellation of Taurus. (The symbol for Pluto is a combination of P and L. It didn't exist in classical, mythological times, so one had to invent a symbol for it. Uranus still carries the initial of Herschel, its discoverer, that is the letter H.) In 1902 Uranus was in sidereal Scorpio (Fig. 7.3). Then, just a few years ago, actually on January 6, 1966, a conjunction occurred between these two planets. It took place in sidereal Leo.

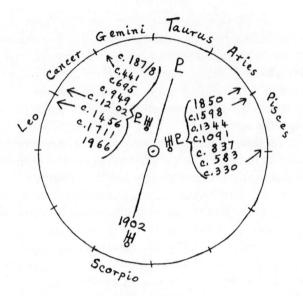

(Figure 7.3)

Now we want to know how this worked out in history. In order to find our way we first go back from the opposition in 1901–02 to the preceding conjunctions and oppositions of the two planets. One conjunction happened in 1850, in sidereal Aries. From there we can move still further back in history, and discover interesting correlations. We thus have actually two rows of conjunctions and oppositions, always at intervals of about 253–254 years:

PLUTO AND URANUS

OPPOSITIONS		CONJUNCTIONS
Sidereal	*Sidereal*	*Sidereal*
1902, Pluto in Taurus,	Uranus in Scorpio	1850, in Aries
c.1648, Pluto in Taurus,	Uranus in Scorpio	c.1598, in Pisces/Aries
c.1395, Pluto in Taurus,	Uranus in Scorpio	c.1344, in Pisces
c.1142, Pluto in Taurus,	Uranus in Scorpio	c.1091, in Pisces
c. 899, Pluto in Taurus,	Uranus in Scorpio	c. 837, in Pisces
		c. 583, in Pisces
		c. 330, in Pisces

(A second row of conjunctions and oppositions converges on different parts of the zodiac. We shall discuss these later.)

All these dates are a wonderful description of historic developments in humanity. In fact they are connected with esoteric Christianity, which eventually developed into Rosicrucian Christianity. We go right back to the years around 330 A.D. There we find remarkable events taking place that were deeply connected with esoteric Christianity. Around that time must have happened what Rudolf Steiner described as the so-called Black Sea meeting. In the preceding century the great Mani, the founder of Manichaeism, which spread far to the East and to the West, had lived and worked. Finally, he went through an extreme experience of martyrdom in Persia. In the following century he gathered, while he was in body-free, spiritual existence, the great initiated leaders of humanity for a decisive meeting. Among them was the Gautama Buddha, who no longer incarnated since he had reached Buddhahood. Also, Scythianos the great initiate of the West, and the great Zarathustra of the East were present. They discussed the situation in which humanity found itself then and decided on what had to be done to safeguard a healthy evolution toward the future. The year 333 A.D. is exactly the center of the so-called Fifth, or Post-Atlantean, Epoch, which started with the Ancient Indian Civilization and will last until the end of the seventh civilization that will terminate this epoch in the "War of All against All" (conclusion in 7893 A.D.). Out of this meeting eventually developed the Christian Mysteries of the Rose Cross, or medieval Rosicrucianism. It was well known in the Middle Ages, among those who were concerned, that the latter was built on those great events at the Black Sea during the fourth century.

The conjunction of c. 583 A.D. was connected with Celtic Christianity, which was a sublime manifestation of esoteric Christianity. In 837 a conjunction took place, and about fifty years later, in 889 A.D., an opposition. Both were associated with the unfolding of esoteric Grail Christianity. It was the century of Parzival. All this was connected with those great decisions that had been taken during the fourth century by the great spiritual leaders of humanity. The conjunction in c. 1091 A.D. was soon followed by the foundation of the Order of the Knights Templar, in 1118. The Templars were, in a spiritual sense, direct descendants of the Grail Knighthood. Another conjunction took place in 1344 A.D., which was accompanied by an opposition fifty years later in 1395. This was also an age of tremendous spiritual decision. At the beginning of the fourteenth century, the Order of the Knights Templar had been extinguished through one of the cruelest and most atrocious deeds in history. The last Grand Master, Jacques de Molay, was burned at the stake in 1314. This was the end of a certain stream of the Christian mysteries. Something had to be done in the spiritual world in order to continue the stream of esoteric Christianity. During the fourteenth century, that mysterious personality whom we know only as Christian Rosenkreutz was born (about 1378), just between the conjunction and the opposition. He lived to a very old age. He is said to have been more than one hundred years old when he died toward the end of the fifteenth century. He was the one who took upon himself the task of carrying esoteric Christianity further by founding the Brotherhood of the Rose Cross, which became spiritually very active in the following centuries.

We come then to the c.1598 conjunction, which was seconded by the opposition of the two planets in c.1648. Those fifty years seem to be connected with the commencement of the mission which the Buddha took upon himself on Mars. Rudolf Steiner revealed that the late Buddha was the great friend of Christian Rosenkreutz. The latter realized the tremendous danger that beset humanity through the decadence that had taken hold of the Mars sphere. Constant and severe wars raged on Mars, through which human souls had to pass when they descended into incarnation. If this situation had carried on, a tremendous split would have occurred in humanity. One part of it would have retired into a purely spiritual, earth-estranged hermit existence. The other part of humanity would have completely submerged itself in the development of materialism. In order to prevent this split, Christian Rosenkreutz implored the late Buddha to go to Mars, to "Christianize," pacify, and harmonize Mars, bringing it back, so to speak, into line with evolution as it was visualized by the divine world. Of course, we do not, on the surface, see great results in our time. Materialistic natural science and the resulting technology seem to be ever stronger, particularly in the last few decades. However, we must not forget that such an impulse needs time for realization. Christianity itself is an example. It is now nearly 2,000 years since Christianity was inaugurated on this planet. And yet we don't seem to see too many results. On the other hand, at present, during the last third of the twentieth century, a surging tide appears to be rising against the prevailing materialism, though it is often misunderstood and misguided.

Finally, we come to the conjunction in 1850 and the opposition in 1902. During that interval another great decision took place. This time it was Rudolf Steiner who took upon himself the task of inaugurating a "Science of the Spirit," Anthroposophy. Thus he offered to humanity the means that are necessary to move toward the future, toward a very practical and yet spiritual-cosmic realization of Christianity.

Now we shall look at the second stream of events, which is associated with the conjunction in 1966. The ancestral conjunctions and oppositions happened as follows:

PLUTO AND URANUS

OPPOSITIONS		CONJUNCTIONS
Sidereal	*Sidereal*	*Sidereal*
		1966, in Leo
c.1793, Pluto in Aquarius,	Uranus in Leo	c.1711, in Leo
c.1539, Pluto in Capr./Aquar.,	Uranus in Cancer/ Leo	c.1456, in Leo
c.1286, Pluto in Capricorn,	Uranus in Cancer	c.1202, in Cancer
c.1031, Pluto in Capricorn,	Uranus in Cancer	c. 949, in Cancer
c. 777, Pluto in Sagittarius,	Uranus in Gemini	c. 695, in Cancer
c. 522, Pluto in Sagittarius,	Uranus in Gemini	c. 441, in Gemini
c. 267, Pluto in Scorpio,	Uranus in Taurus	c.187/8, in Gemini

After a series of earlier events which we cannot discuss here, we come to the year 1202, which is a conjunction. The years around 1202 are marked by two totally different developments. Altogether it seems that the conjunctions of this order are, as a rule, accompanied by contradictory historic events. 1202 saw the culmination of the movement of the Troubadours. It was the time of Wolfram von Eschenbach, of Robert de Boron, of other troubadours who wrote about the Holy Grail and brought it thereby to the knowledge of later humanity. Then we come to

1456, which was also a conjunction. A little over two years later, in 1459, the "Chymical Wedding of Christian Rosenkreutz" took place. The document that speaks of it—one of the fundamental writings of medieval Rosicrucianism—bears the addition "Anno 1459." On the other hand quite different things happened. About 1206, which is very close to the first conjunction (we must, of course, imagine that these planets are moving very slowly and stay close to each other for a long time), developments occurred in the interior of Asia. It was the time when Temuchin was elected by the Kurultai (the assembly of the nobles of his territory) to be Genghis Khan, the perfect warrior and the ruler of the world. In Europe, the Troubadours worked in remote corners collecting and preserving the story of the Holy Grail, of Parzival, of esoteric Christianity. And in the East the counterblow was prepared. The Mongols under Genghis Khan and his successors conquered a great part of Asia and even broke into central Europe. They came as far as Liegnitz in Silesia. A European army that opposed them was actually completely defeated at the battle of Liegnitz in 1241. The road into central Europe and to the West was wide open to the Mongols. Then something happened for which history has no real explanation. The invaders turned back and disappeared toward the East. One suggestion is that the reigning khan at home, the successor of Genghis Khan, had died and that it was necessary to elect a new khan. A legendary explanation is that although the armies of the Europeans had been defeated and there was no longer any resistance, a few nuns in a monastery went into deep meditation and prayer. Thereby a wall of spiritual resistance was erected against this invasion. A few years prior to the conjunction in 1456, in 1453, there came again an invasion from the East into Europe. The Turks then took Constantinople and thereby established their rule in Europe. Earlier, during the seventh and eighth centuries, something similar happened. In 711. shortly after the conjunction in 695, the Moslem Moors from Africa crossed the Straits of Gibraltar and defeated the Visigoths in Spain. Thus, step-by-step, they took possession of the Iberian Peninsula.

We see how there were constant attempts made from the Asiatic continent, from an Old World, to destroy the integrity of

Europe. In the case of the Mongolians, it was even the oldest, the last shadows of Atlantis, which manifested themselves through that race. From the Moslem world came first the grip via the north of Africa into Europe, via Spain. Several centuries later, in 1453, came the blows on the eastern flank, via Asia Minor into Europe, by the taking of Constantinople and most of the Balkan Peninsula.

In 1711–12 the last but one of these great conjunctions took place. It was accompanied in history by the birth of J.J. Rousseau (June 28, 1712). On the other hand it also had a deep connection with J.W. Goethe. Goethe was born thirty-seven years after this event, on August 28, 1749. Thirty-seven years corresponds to two cycles of the moon nodes. Thus the nodes returned at incarnation to the position that they had occupied in 1711–12. The fact that he had a deep connection with 1711, even while he was still unborn, is demonstrated by the events in his life eighty-two years later, in 1793, when the succeeding opposition of Uranus and Pluto took place. Around that time Goethe wrote down the "Legend of the Green Snake and the Beautiful Lily." This is essentially a Rosicrucian legend. Goethe had indeed a deep connection with the Rosicrucian impulse, which is also evident in his poem "The Secrets." The legend speaks of the bridge that is to be built between the land of the Beautiful Lily and the mortals who live across the river, on the earthly plane. Finally the bridge is established by the sacrifice of the Green Snake. By spanning the gap above the river with her own body, she reconnects the land of the Spirit with the realm of the Earth. Thereby people can again find access to the world of the spirit. This was genuine Rosicrucianism. It is already expressed in the "Chymical Wedding of Christian Rosenkreutz, anno 1459." In a deeper sense it is a description of how a western humanity can again find its way to a direct cognition of the spiritual world.* The black cross stands as a signum for all that which appertains to a physical-material world, to that world in which we human beings, and in the end everything in nature, experience death and decay. But death

*The principal "symbol" of Rosicrucianism, the Rose Cross, is a witness to this and a "road sign" to its achievement.

can, and must, become the portal to resurrection. The attainment and power of resurrection is signified by the glowing red roses on the cross. They are symbols of life, even eternal life. Thus is Rosicrucianism, indeed, esoteric Christianity. Goethe took it upon himself to call upon humanity through his legend, through a kind of fairy tale. He thereby described the great task and responsibility of modern humanity, to build again bridges from the world of the visible to the world of the invisible, to cognition of and living with spiritual reality. The events in Goethe's life around 1793 were indeed the result of the earlier occurrences, accompanied by Uranus and Pluto, in 1711 and 1456-59.

Next we come to modern times, to the conjunction in 1966. This will be followed by an opposition that will take place around 2047. It will fall into the twelfth seven-year period after 1966. We can expect that the approach of this event will be a most crucial time, which the experiences and the evidence we have so far confirm. Of course, our question must be, Did the conjunction in 1966 have any connection with the Rosicrucian impulse? Our impression is that the latter again took a decisive step forward in preparing the future of humanity, even preparing the Aquarian Age, which, in our opinion, is not yet. Rudolf Steiner painted in the smaller cupola of the first Goetheanum a remarkable picture that seems to be significant in just this context (reproduced in *Wege zu einem neuen Baustil,* by Steiner, published 1926). At the lower part we see a representation of Slav humanity, which will eventually lead the Aquarian Age of civilization. Sideways above it appears a modified Rose Cross. The seven roses are replaced by seven pentagram-stars. What does this signify?

The rose is built on the principle of five petals, or multiples of five. They form a pentagon (Fig. 7.4). The pentagon, like the pentagram, is a cosmic symbol. It is associated with the movements and rhythm of Venus, the planet of love and compassion. (The conjunctions of Venus with the sun inscribe a pentagon-pentagram into the space around the Earth. See Chapter I.) In a true sense it is even connected with the Gautama Buddha, who brought to humanity the teaching of love and compassion as preparation for the Christ Event. This can be verified by a historic cosmography.

The symbol of the roses on the Rose
Cross is the sign of the power of resurrection, as we said earlier. They replace
the corpse on the crucifix and thereby
can lead over from the Good Friday
experience to Easter Sunday morning,
to the experience of the Resurrection.
(In this context it is essential to study
Rudolf Steiner's extensive description of
these matters in his *Occult Science*, ch. 5.) The Pentagram

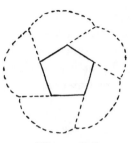

(Figure 7.4)

Cross can lift this experience still a stage higher. The plant—in
this case, the rose—is a pure and undeviated manifestation of
cosmic forces working on Earth. With some imagination one
could say: the rose's great ideal is the planet Venus, or rather
the sphere of the latter. It works in the cosmos on the foundation of the pentagon-pentagram. And we have the impression
that Rudolf Steiner replaced the roses in that painting by pentagram-stars because he wanted to intimate that future
humanity, eventually the humanity of the Aquarian Age, must
rise to this Pentagram Cross, in order to realize the cosmic
impulse of love and compassion on Golgotha in social deeds.
And in the midst of the events of the last third of the twentieth
century, we seem to experience that this dramatic development
is already with us, in the very depths and secrets of modern
humanhood. Of course, we also realize that the antipowers are
doing their utmost in trying to prevent this impulse from
entering human consciousness. Great spiritual strength and
perseverance will be needed by the individual if he or she
decides to join in with what appears to be a new step forward
inaugurated by genuine Rosicrucianism. In fact, this implies
that humans need no more imagine themselves to be helpless
creatures, constantly subdued and dominated by the stars.
They can learn to become spiritually free "cooperators" with
the cosmos. This was one of the last, great deeds of Rudolf
Steiner, to point out ways and mean by which humanity can
move toward a future in which it will take the mighty ingredients of the cosmos, fulfill them spiritually during a life's experience, and eventually hand them back to the cosmos, enriched
and renewed. Humanity, seen from a long-range view of cosmic

evolution, will eventually help to build a new cosmos. And present humanity seems to be called upon to enter the very first stages of such responsibilities.

To such a level of realization a modern cosmology and cosmography can, and should, raise itself.

8.

Turning Points in History—II

AS A CONCLUSION to this week's work we shall try to discern the significance of the present century and its possible projection into the future. Last night we mentioned one of the major events in the beginning of this century, which somehow seemed to signify its character. To this we shall add two more features, one in the beginning of the century, which almost coincided with that opposition of Pluto and Uranus of which we heard, and another one exactly at the end of the century. In 1901 a Great Conjunction of Saturn and Jupiter took place (Fig. 8.1).

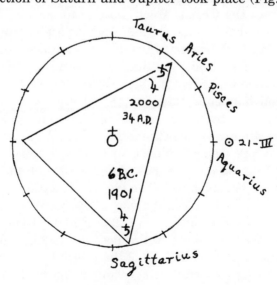

(Figure 8.1)

This happened in the constellation of Sagittarius. In 2000 another such event, also a conjunction of Saturn and Jupiter, will take place somewhere in the transition from the sidereal constellation of Aries to that of Taurus. These are two of the three possible conjunctions that form the Great Triangle, which has been mentioned earlier. It is one of the hands on the cosmic clock that we need in order to gauge time, in a cosmic as well as in a historical sense. Each one of these conjunctions recurs at intervals of sixty years. For instance, the one in 1901 returned in 1961. The one to come in 2000 actually happened in 1940–41. Furthermore, we notice that the one in 1961 was further forward in the ecliptic than its predecessor in 1901. Likewise the conjunction of 1940 was further back because it preceded the one in 2000. Thus the triangle turns very slowly forward, like the hand that we use to gauge time.

We can also go back in history and find that the Great Conjunction of 1901 leads us right back to a similar event in 6 B.C., to an ancestor of the one in 1901. Because of the forward movement of the triangle we find it then, in 6 B.C., in the constellation of Pisces. This was the Great Conjunction, the conference of Saturn and Jupiter, which announced the "Fulfillment of Time," the birth of Jesus, who was eventually to become the vehicle of the Christ. Much more is to be said about this. We have worked out the details and report on them in *Cosmic Christianity*. The event in 2000 A.D. can also lead us back into the past. Eventually we find an ancestor in the year 34 B.C., that is, one year after the Mystery of Golgotha. The Great Conjunction happened then in the sidereal constellation of Leo. From Leo in 34 A.D. it moved all through the zodiac, and by 2000 it will arrive near the ingress from Aries into Taurus. The year 34 A.D. seems to have coincided with the conversion of St. Paul. Thus we have in these conjunctions an inspiring pattern of events. The opening of the twentieth century is accompanied in the cosmos by a conjunction whose ancestor was associated with the "Annunciation" in 6 B.C. Rudolf Steiner realized this in the inauguration of Anthroposophy, in his announcement of Anthroposophy. At the end of the present century would stand, as far as we can gauge, the cosmic memory of 34 A.D., the conversion of St. Paul. These are, from one angle, the foundations of the present century. It seems to be important to bear this in mind. However, how is all this connected with the future? How can this possibly guide humanity?

In order to find an answer we must look at the present century from a broader aspect. We say we are at present still in the Age of Pisces, and we are moving very slowly toward the Age of Aquarius. What do we mean by this assertion? The beginning of the year is, in a cosmic sense, not on the first of January. The actual astronomical commencement of the year should be seen as the time of the vernal equinox, when the Sun is exactly at the crossing point of the celestial equator and ecliptic, about March 21st–22nd. We know that the fixing of Easter is closely associated with this: after the commencement of spring, the vernal equinox, and after the first Full Moon following it, Easter will be on the succeeding Sunday. Thus the position of the Sun at the vernal point should be regarded as the moment of the commencement of the year of 365 days. It is much more than only a convenient way of organizing the yearly calendar.

At present we would see, with efficient astronomical instruments, behind that Sun around the 21st of March, the fixed stars of the constellation of Pisces. This was not so 2,000 years ago. Then the Sun was standing in front of the constellation Aries at the time of the vernal equinox. In fact, it was just moving over from Aries to Pisces. This is called the movement of "precession" in astronomical language. The fixed star constellations shift very slowly forward in the ecliptic, away from the vernal equinox. Thus the constellation of Pisces has moved into the place of Aries, and soon the constellation of Aquarius will replace Pisces. In about 600 years' time the star Eta, in the urn of Aquarius, will be seen behind the Sun around the 21st of March.

However, the Sun will actually be seen much earlier in that constellation of Aquarius. In about 400 years' time the Sun of the vernal equinox will already be above the region of the water that the Waterman pours out of his pitcher or vessel. This stream of water is indicated by a number of very faint fixed stars (Fig. 8.2). In about 2369 A.D. the vernal point will have moved to a point above this stream of water. One of the Fishes will still be above it. Thus a kind of double aspect is indicated,

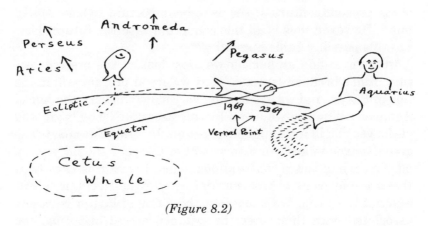

(Figure 8.2)

which we shall have to face then in 400 years' time. At present some people seem to be in a great hurry and maintain that the vernal point is already in the Waterman. However, this contradicts the astronomical facts, on the basis of the generally acknowledged maps of the fixed-star heavens. Our contention that the ingress will happen only in approximately 400 years' time has been confirmed by Cyril Fagan, an astrologer and astronomer, who is very precise and correct in his calculations.

Thus we shall then, during the twenty-fourth century, have a kind of three-cornered proposition in the cosmos: There will be the waters of the Waterman below the ecliptic, above it the western Fish, and still higher the constellation of Pegasus. This situation will have to be prepared by humanity, in the sense that a future, new cosmology will require that we "speak" actively and creatively to the stars. The cosmos in which we live has exhausted itself in the past stages of creation to the point where it only reflects and remembers the deeds of the divine world in the past like those in a kind of great chronicle. Now it has become our responsibility to rise to such levels of moral imagination and intuitive thinking that, eventually, we will be able to contribute to the reawakening and even reenlivening of the cosmos. It is the still very small starting point of a new creation. This is what St. Paul meant when he spoke of the First Adam, the first creation, and the Second Adam, the creation that can commence when we take the Christ Impulse into our own being.

We can well ask, How can we be expected to prepare for all this? The danger is that nothing may happen in humanity. Then these Aquarius forces would no longer create, but destroy. Aquarius has two sides in mythology: one is connected with floods, the other with the fertilization of the earth. In Chaldea it was associated with the mighty flood described in the epic of Gilgamesh. In Grecian mythology Aquarius was identified with Deucalion and his wife, Pyrrha, the only survivors of a great flood in Thessaly. Therefore humanity itself must make a start from Moral Imagination and Intuitive Thinking, in order to prepare. This can only be done by fulfilling first the tasks of the Piscean Age. The Piscean Age is not yet at an end. In the sense of spiritual culture it has barely begun. We cannot hope to enter the Aquarian Age safely before having comprehended the tasks of the Piscean Age. What are these tasks? We can no longer read the requirements and expectations in the stars. The doing must come from us ourselves, and can come from us only. We find above Pisces (Fig. 8.2) the constellation of Andromeda. She is chained to a rock and is threatened by a terrible monster down below Pisces. This is Cetus, the Whale. It is a rather formless, watery creature of the great sea below this part of the zodiac. Andromeda's mother, the queen Cassiopeia, had angered the God of the sea, and in revenge he sent that monster, Cetus, to the kingdom. It ravaged the region terribly. An oracle said thatonly one thing would help to free the country from this calamity, the sacrifice of Andromeda. So, she was exposed to Cetus, in order to be eaten up by it. However, just at the moment when Cetus approached there came help from above. Perseus came flying home from an expedition on which he had killed another monster in the far north, Medusa. Cetus and Medusa belonged to a generation of monsters who were once beautiful beings. But getting old, surviving into a later age, they became ugly and dangerous. Medusa was capable of transforming into rock everything that looked into her eyes. Perseus, however, was a child of the Age of Aries, that is the Greco-Latin Age. He had learned to think, a capacity that manifested so wonderfully in the later philosophers of Greece. Thus Perseus did not approach Medusa directly. He turned round and observed the face of Medusa in his shield, which was polished and blank like a mirror. He approached her without being converted into a rock, and cut off her head. On the way home he saw at once the danger to which Andromeda was exposed. Instantly he took out of his bag the head of the Medusa and held it in front of the approaching Cetus. In one moment the latter was transformed into a huge rock. Indeed, in later times a big black rock, off the coast of Phoenicia, was supposed to have been once upon a time the body of Cetus.

These myths contain truths of evolutionary import. What does it mean that Andromeda is chained to a rock and sacrificed to a monster? This monster was a sea creature, formless, gigantic, and dangerous. Who rescued her? It was Perseus, who had developed the power of thinking. Down there, in Cetus, thinking was not represented. Cetus was something that had become atavistic and formless, something that was connected with the sea. It was an ancient, etheric clairvoyance that had not gone along with the evolution of human consciousness and therefore was dangerous, though intriguingly voluminous. It threatened Andromeda, who must be regarded as an image of the human soul. There is a message that especially concerns our present age. Andromeda is above the Fishes, the constellation that inspires the present vernal equinox. There is one star on the forehead of Andromeda, just in this region of the Jupiter chakra, which is according to esoteric conceptions the "two petalled lotus," an organ of higher perception described in Rudolf Steiner's *Knowledge of the Higher Worlds*. Andromeda has this star in common with Pegasus. It is also in one of the wings of Pegasus, and it looks on classical star maps as if Pegasus was rising out of the head of Andromeda. The myth expressed in this is: Andromeda, the soul of humanity particularly concerned with the Age of Pisces, has the impulse to develop the capacity associated with the "two-petalled lotus." It is the raising of thinking to a higher level, to a "clair-thinking" as I might tentatively call it, clair-thinking suggesting the modern equivalent of clairvoyance. In the sense of the image of Pegasus this is "winged intelligence," or cosmic intelligence.

Andromeda was rescued by the newly acquired capacity of the Aries Age, the power of thinking. The ability to think with the brain had to be evolved first before progress to "clair-think-ing" will be possible. And then the atavisms of the Cetus-nature can be overcome. This is the task of the present Age of Pisces. It is not too pessimistic to say that we have not yet achieved that aim. Rudolf Steiner brought the message to present humanity in 1893 when he published the *Philosophy of Spiritual Activity*. This is the foundation and starting point from which we can proceed to evolve that Moral Imagination and Intuitive Thinking that modern humanity needs so very badly. When we look around, everywhere, in all fields of human life, we find the same phenomena: we are working with worn-out concepts, with a thinking that is coming from the past, in many instances still derived from Roman times, partic-ularly when it comes to political concepts, to concepts supposed to form human community. Therefore so many things go wrong, simply because we have no constructive, new, indepen-dent ideas. Only when we have developed these new approaches to thinking can we hope to move safely into the Age of Aquarius.

What will the Age of Aquarius require? Here we have to con-sider several things. The civilization of Aquarius will need to handle and evolve the Saturn forces, that is, the forces and power of "Gnosis" in a cognitional sense. Why should this be so? The planets have their definite "domiciles" in the zodiac. For instance, Jupiter is at home in Pisces. This means that the par-ticular quality of this planet must be developed during the present age. An example of a positive development of such qualities was Goethe. He was deeply connected with the con-stellation of Pisces and Jupiter at the moment of his incarna-tion. The thinking that Goethe developed, how he evolved his relationship to the world, stands as witness. In the sphere of science it was the "archetypal plant." In other realms, for instance in his poetry, he created an archetype of the modern human being in the image of his "Faust." Goethe was truly a member of the Age of Pisces, carrying Jupiter to a certain point of fulfillment with all that he did.

However, the times of the presence of the vernal equinox in Aquarius, and also in Capricorn, still further in the future, will see the last civilizations of this present cycle of the seven civilizations of the Post-Atlantean Epoch. Both will be challenged by Saturn as the planet who has his domicile in these constellations. In order to comprehend the challenge of Saturn, we would have to go right back to "Ancient Saturn" (see R. Steiner's *Occult Science*). Of course, we do not suggest that on Ancient Saturn there already existed a Capricorn or a Waterman. However, in their places there were dwelling hierarchical, divine beings who inaugurated the evolution of humanity and of the world. For instance, in the Capricorn of today, we would read the story of spiritual beings creating the very first etheric germs of the sense organs. And Aquarius tells a similar, though slightly different, story. As this constellation appears in the heavens nowadays it is only the chronicle of those events in the very dim past. Thus is Saturn, which is a memory-replica of "Ancient Saturn," connected with Aquarius. We will have to evolve so that we can eventually employ these forces as our own. In order to prepare for this, we have to cultivate now, in this age, the Jupiter forces first. Then we will be able and can proceed to develop the Saturn forces at some future time.

What will actually be the task of the Waterman Age, then? Indian mythology connects Waterman with the God Varuna. According to the description, he is like an equal, or pre-image, of Ahura Mazdao of later Persian civilization. Ahura Mazdao of Persian times was the "Aura of the Sun," not the Sun that we see in the sky, but its aura, actually all the Sun activity that penetrates the totality of the solar system. This Light Being in the heavens is opposed by Ahriman, who dwells deep down in the center of the Earth. He is the "spirit of darkness." Persian civilization was built on this great imagination. Zarathustra had thus inaugurated a civilization that worked in all fields of life with these tremendous concepts. For instance, organized agriculture was founded on it. The soil of the Earth was opened up with the plough, in order to let the light of the cosmos, of Ahura Mazdao, stream into the dark earth, so that the plants which, constantly overcome the darkness of the earth, could grow. They reach with their roots into the soil and grow up with their stems, leaves, and flowers toward heaven. Thus they imitate and manifest the heavens on Earth. The

God Varuna of Indian civilization is described by A.L. Basham in his book *The Wonder that was India* as follows: "Varuna is sitting in a great palace in the heavens. He is the guardian of Rita. (In Indian writing this is spelled Rta.) Rita is the Order of the Great Cosmos. All the great rhythms of the cosmos, day and night, down to the change of the seasons are carried and caused by this Rita, the Cosmic Order. Also humanity must live according to Rita. In later India non-Rita, the opposite of Rita, or Cosmic Order, became the commonest word for untruth, sin, and anti-order."

Varuna, the great Guardian of the Cosmic Rita, rides across the heavens on a steed. This latter is in Indian mythology associated with Capricorn, the Makara, sometimes depicted as a sea monster. Of course, these are ancient Imaginations, and we in our age must find new ways and learn to speak again a language of the stars that modern humanity will understand. We must learn to comprehend the meaning of Rita in a modern, even scientific, sense. It is the world of the etheric forces, the formative forces, working from the cosmos into Earth existence and creating life in nature and in the human being. This will be the great task of the Waterman Age. Cosmic "water" will have to be handled very consciously, not only earthly water, though this is certainly connected with the life forces in nature. This will be possible only on the basis of a creative, of an intuitive, thinking. The handling of the evening and morning forces, which we mentioned earlier and which is done in Europe even for practical purposes, is the very first step in this direction. Human beings of that future will learn to receive the cosmic forces into their hands and work with them, bringing growth and healing into the kingdoms of nature, including the kingdom of humanity.

In order to accomplish these great tasks, we have to learn one thing first, which is the potential capacity that we see indicated in the cosmos by the image of Andromeda. We must attain that Intuitive Thinking and Moral Imagination of which we spoke earlier, and which is connected with the development of the two-petalled lotus. The present age is already moving toward this. That inauguratory Great Conjunction of Saturn and Jupiter in the year 1901 in the constellation of Sagittarius suggests it. Just as we can study the past history of this Great Conjunction, so can we also go forward, toward the future. Thus we discover that in a few hundred years' time, this event will take place in Aquarius. First, in the course of the next century, it will move into Capricorn, and after that into Aquarius. Thus our present century stands as one of the more significant ones since the time of Christ. We are in a position at present where the Great Conjunction of the "Annunciation," descended from its ancestor of 6 B.C., has reached a certain climax. Rudolf Steiner obviously perceived this clearly at the beginning of the century. He did "speak to the stars," and his speech was Anthroposophy. In 1901 when that conjunction took place in the heavens, he commenced with two extensive cycles of lectures: *From Buddha to Christ* (given in a group called *Die Kommenden*), and *Christianity as Mystical Fact* (given at the theosophical library in Berlin). Both are now published in book form. Thus we must expect that this conjunction will go through the coming centuries, standing in the heavens like a perpetual challenge, like the external signs of conferences between the spiritual beings who are associated with Saturn and Jupiter. And they will say: We look toward humanity— what will this humanity do? How will this humanity react to the challenges? Our answer can only be the relentless and patient development of what Rudolf Steiner entrusted to humanity during the first quarter of the twentieth century.

One of the first books written by Rudolf Steiner was *Knowledge of the Higher Worlds*. About twelve years before that he published *The Philosophy of Spiritual Activity*. His intent with these publications and all the activities that followed was to show ways that would eventually lead to a scientific breakthrough to spiritual perception, to a new awareness of the higher worlds, which are closed to the physical senses. Thus can modern humanity rise to new levels of experience and of capacities by which it can master its life on this planet in a constructive and positive fashion. By no means did Rudolf Steiner at any time want to satisfy personal curiosity, or even desires for entertainment. He did everything from the realization that the whole cosmos is looking down on present humanity and expecting ideas and deeds that will bring the world forward, to continue the great work of creation that the hierarchies began long ago. Thus indeed has the significance of the twentieth century been made manifest by Rudolf Steiner's life-work: a humanity standing at the crossroads of spiritual freedom, a freedom that it must learn to employ in order, in the however distant future, to rise to the ranks of the creative powers of the universe.

PART TWO

Cosmic Christianity

The Stars during the Three Years of Christ's
Ministry and Practical Viewpoints
with Regard to Evolution

The revised content of eight lectures given at
Hawkwood College, Stroud, Gloucestershire, England
August 24–31, 1969.

These were not, strictly speaking, lectures in the usual sense.
They were conducted as "work" sessions, in which
the lecturer gave leading ideas and practical suggestions,
which all those present then worked out together.
Thus the meetings were a "give and take" in which the whole
audience was actively involved.

1.

The Movement of Saturn during the Three Years

WE DECIDED TO WORK over the so-called "Three Years" and in a wider sense over the thirty-three years, from the birth of Jesus up to the moment of Golgotha. Why do we want to do this? There are many reasons that one can point to in order to justify such an intensive study.

First, I would like to quote the opinion of Rudolf Steiner, published by the periodical *Blätter für Anthroposophie,* 4 Jahrgang, No. 6: "If the correlation between the Earth and the extraterrestrial world—that is, the cosmic world—is not comprehended again on a level of spiritual understanding, then the Mystery of Golgotha cannot live on, cannot survive into the future." What does this mean? It simply means that in order to make the Mystery of Golgotha live on into the future, it will be necessary to grasp, on a spiritual level, the interconnection between the cosmic and the earthly and human worlds, for instance, during the Three Years. Another point is this: we must gradually move to the level that St. Paul, nearly 2000 years ago, had reached in the moment when he could say, "Christ in me—not I myself, but Christ in me." To come to such an experience needs a tremendous inner development. Humanity in our present age is called upon to move toward that position. This again needs the knowledge of cosmic correlations.

Another point is that in order to rise to the heights of Christianity, to the esoteric meaning of the Christ events, we must take into account the cosmic events, and inner experience can thus increasingly become a reality with regard to the Christ events. This will become more and more evident. One can even say that those events which took place nearly 2,000 years ago will come back to humanity, even as a "Presence." One day people will realize not only that the Christ events did take place 2,000 years ago but also that those events can be here and now with humanity, but in a new fashion, not in the physical but on a higher plane, on a plane that is accessible only to spiritual experience. A new clairvoyance will arise in humanity. This is a very profound perspective. You know yourself that all religion finds itself at present in a very difficult situation. There are so many problems. We have heard it proclaimed publicly, even by ministers of religion, that God is dead. You see, if this should be widely accepted, it would mean the end, not only of religion in general but of Christianity as well.

Also, you know that the historicity of those events, about 2,000 years ago, is by no means established as a fact in modern history. This branch of science is still in doubt about them. These doubts made it possible for a professor at a German university at the beginning of this century to suggest in all sincerity that the Gospels are nothing but fictional stories concocted by clever astrologers who knew the events that took place at the time of Christ in order to gain insight into their probability, and that on the foundation of this knowledge they concocted the Gospels. You see, this is a very serious matter and we ought to occupy ourselves with the events closely in order to get insight into the probability of the events. For this we must rise to the true heights of Christianity, to the esoteric meaning of the Christ events. We must take into account the cosmic events.

Finally, I should like to point out that the humanity of the future will move more and more toward a realization, toward a preparation for the realization, of what is sometimes called the "phantom body," that is, the "resurrection body." This can be done only by a spiritual knowledge of the correlations between the cosmos and the Earth. Take only the following facts: We so happily rely on nature. Each time a human being is born on this planet, nature provides us with a body. We take it for granted that we are endowed with a physical-material body, but are we aware of how this came about? It is given to us and we take it as a matter of course. We can know, of course, and

Anthroposophy gives us all the information we need, that the human being, together with the spiritual hierarchies, builds up the form body, the spiritual form body, in the cosmic world long before incarnation. We receive and bring it down from the cosmos. The zodiac is the external expression of the spiritual forces of that form body, that body that had been designed, as it were, in the dim past by the divine world. This we receive, and with this we force the earthly matter, which is offered to us at the moment of conception, into a human form. Without that spiritual form body we could not do it, and this is bestowed on us. There will come times when this will cease, when we will be called upon more and more to do it ourselves, to do it as a self. This we will be able to accomplish only if we have the corresponding insight. First, we have to prepare ourselves by creating a knowledge of this correlation between the great cosmos and the incarnation in Earth-matter. This preparation will have to include the planets—a knowledge of the functions of the planets. It will become more and more necessary. This cosmic form body, this eternal archetype that really makes the human form, was founded in the very beginning of all evolution.

In the course of human history, this cosmic form body was lost by the human being. Through the "Fall From Paradise," the human being became deeply involved in corruptible matter. Since that time, whenever a human being incarnates, the incorruptible cosmic form body is torn down into the domain where death rules. Thus the human being is irresistibly inclined to take the corruptible body for the only reality. However, the incorruptible body was rescued by the Deed of Christ. This is the resurrection body. This is that body in which Christ appeared to the disciples. It is the form of which He could then even say to Thomas: "Put your fingers into the stigmata, so that you can convince yourself of my Presence."

We have to prepare for this realization as we move toward the future. In order to take the first steps of preparation, we must get a glimpse, we must start to acquire an insight and a knowledge, of the correlation between the cosmos and humanity, as it was reestablished by Christ.

Now let us go to work. We shall start with the planets, and simply look at what the planets did during those three years.

There is straightaway a problem that is not easy to solve. It is the so-called "Three Years." The question is what is meant by the Three Years chronologically. Some people have formed the opinion that it was three and one third years, that the Baptism took place, if the traditional date is accepted, on the 6th of January, the day of Epiphany, of the year 30. From there we would count three and one third years up to April 3, 33. Another view is that it was only two and one third years. We hear in the Gospel of St. Luke that Jesus was thirty years old when he went down to the Jordan in order to be baptized. That would have been the moment when the cosmic Christ-Being entered the body of Jesus. I have investigated both possibilities very closely, and I myself have come to the conclusion, as far as the cosmic evidence goes, that the version of two and one-third years is more realistic than the other. Jesus would then indeed have been thirty years old, just a few days prior to the 6th of January, the day of Epiphany of the year 31.

We take first Saturn. What is Saturn doing? Saturn is, in a broad sense, an expression; it is not the beings involved themselves, but it is the cosmic expression of the Father Forces in the universe. They inaugurated, in the dim past, evolution. We have to go back as far as Ancient Saturn according to Rudolf Steiner's *Occult Science* in order to get an idea of the workings of the Father Hierarchy. There exists an intimate relationship between the present planet Saturn and Ancient Saturn. In a sense, the present Saturn with its rings is a perfect image of Ancient Saturn. It is only a smaller, condensed image of the original Ancient Saturn. The latter was as big as the sphere, that is, the volume of space that is contained in the orbit of Saturn. Around that physically very subtle "body"—it consisted only of warmth—stood the divine hierarchies in a big circle, like the rings of present Saturn. The living memory of this organized, so to speak, the present planet. Thus our present Saturn is, first of all, an image of that Ancient Saturn. And in this sense it is an image of the line of divine intention, the focus that goes through all evolution and that will go through into the far future. It is a line indeed, and this is present in every human form as the power of human uprightness. This element of uprightness in the human form is, in a sense, a kind of metamorphosis of that long line of divine will from the dim past into the far future. One can regard it as the ordination of evolution by the Father Forces. Thus would Saturn know the dim past, and project, as it were, the past into the future. We can see in it an expression of sublime historic conscience. The movement from the past toward the future is wonderfully expressed in the form of the human skeleton. First, there is from the past the enclosure of the skull. It encloses the brain and has the tendency to contract and to concentrate. Then the skeleton spreads through the body. As much as Saturn guards and hems in the brain, so does it change its character as we move into the body. There is no longer any brain, though still a spinal cord, nerves, and the muscles of the body. The latter are outside and around the bones of the limbs. Indeed, a tremendous metamorphosis has taken place here. Saturn started in the skull and evolved step by step right into the skeleton of the limbs and down into the toes.

Now this Saturn moved during the Three Years from the constellation of Gemini, the Twins, through the constellation of Cancer. About the time of Whitsun 33 A.D., it entered the constellation of Leo. (See Figure 1.1, taken from the geocentric viewpoint.) What does this diagram tell us? What speaks in this moment from the far spaces of the cosmos? We said that it

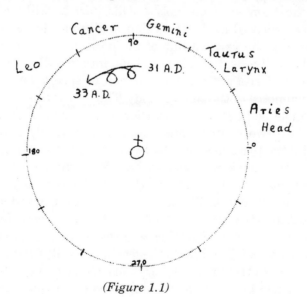

(Figure 1.1)

is a representation, a picture or image, of the Father Forces. What did that Father World pronounce in that moment? This world speaks continually through the external medium of the stars. But we are no longer capable of listening. Otherwise we could by Moral Imagination make something of these things, and the most wonderful inspirations could rise up in us. This we can learn again, though in entirely new ways. Concerning the Christ events, we imagine that there it was expressed and given an entirely new significance on the Earth through the words and deeds of Christ.

First, we have to go a bit further back from Cancer to the constellations of Aries and Taurus. What do these constellations of the zodiac represent with regard to the archetypes of the human form? The cosmic spiritual form body is represented in the constellations, the twelve regions of the zodiac. There we find the image of the totality of the human form as it was ordained by the divine world in the beginning. Aries is the image of the head. Taurus is the archetype of the larynx, the whole region of the larynx, including the canals leading up into the middle ear. It not only concerns speech but also hearing. The canals, or Eustachian tubes, are like horns going up to the middle ear. The whole organism appears almost literally like a bull. Further back you find the Ram with his horns. This part is still, so to speak, faintly indicated in the lines that lead from the horizontal eyebrows into the vertical line of the nose. (Υ) Now in Gemini, that is the Twins, there is a division into two. This physiological tendency descended from the region of speech and sound into the symmetry of the human body. Thus it continues throughout the whole body. It started earlier in the head as the two ears, two eyes, two nostrils, and so forth. Eventually it becomes in Gemini the two arms and the two feet. In the Bull-larynx region the word is spoken, and now here in Gemini it is realized as movement and deed. The next step, Cancer, or Crab, is the house of the chest. The chest is something wonderful if one understands it in the sense of a spiritual anatomy. It is built of twelve pairs of ribs. This "house" is built according to the cosmic order of twelve. The human heart dwells in it, where life is centered, where the breathing organism is at home, and so forth. Then comes Leo. Leo is the archetype of all that which dwells within the "house," where the circulation radiates out into the totality of the body. So, what does it mean when Saturn is in that position? It says, starting out from the cosmic image of the head, that Saturn has now moved through the region of the cosmic Word, the Logos, and has entered the region of Gemini, the Deed. In that moment the Incarnation takes place. It is done, the "Great Event" is done. Christ enters the body of Jesus. Saturn was, during most of the time of the Three Years, in Cancer. He, the Christ, dwelt in the "house," the house of the originally divine human form, from which were imparted the new impulses of evolution to Earth existence. When Saturn enters Leo we see something like a mighty communication of cosmic heart forces to the community of the first Christians. At the Whitsun Event they all experienced, as they were sitting there, something of the universality of the Christ impulse. They partook in it as a kind of cosmic communion. This was a tremendous, new manifestation of the meaning of Leo. The ancient symbol of Leo is this: ♌ you start out from the heart but then you move out to the periphery, for instance, to the periphery of the body, via the stream of the blood circulation. This happened, in a spiritual sense, to the whole community of first Christians who were present at the first Whitsun.

Ancient mythology can tell us wonderful stories of these constellations. Norse mythology in particular gives us a wonderful impression of the constellations. It speaks of Gemini, for instance, in connection with the death of Baldur, the God of Light and of the Sun. He is killed by his own brother, Hoedur. The Evil One, Loki, had a hand in it. Odin once wandered through the world and he heard a prophecy that Baldur would soon be killed. He came home to Valhalla in great distress. The gods discussed the matter among themselves and decided to send messengers into all the world to extract promises from all beings that they would not touch Baldur, not kill him. This was done. The messengers came back. Everything had been accomplished to satisfaction. Only one plant was forgotten, the mistletoe. That happened because it did not seem to be a plant of its own; it is attached to other trees and therefore it was forgotten. Loki, the Evil One, knew this. There was now a great festival in Valhalla, and the Gods threw everything they could lay their hands on at Baldur, because they knew nothing could hurt him. Hoedur, his blind brother, did not participate. So Loki crept up to him and asked him, "Why don't you take part in this sport?" The reply was, "I can't see a thing." "I shall help you, I shall guide your hand," said Loki. He pressed an arrow made of mistletoe into his hand and passed him the bow. Hoedur shot, and Baldur fell dead. We see here twins, the God of the Sun and of the Light, and his brother, who lives in darkness because he is blind. This is the blindness that gradually took possession of the whole of humanity as the gate to the ancient, direct awareness of the spiritual world was closed. And that "blind brother" kills the light. Heaven and Earth were separated. Down in that darkness a humanity developed that became more and more separated from the reality of a divine world. Baldur, the Sun Being of Promise, was in the heights of the spiritual-cosmic world. There grew a tremendous gap.

Now, we can really see what actually did take place in that moment of the Baptism. It seems to be a contradiction at first. But this is just the wonderful fact: the Christ Event healed and redeemed that split between Heaven and Earth. This happened when John the Baptist heard the voice in heaven of the Father: "This is my beloved Son." In the usual Bible translation it says: "This is my beloved Son in whom I am well pleased." Rudolf Steiner says this translation is incorrect, and it actually should be: "This is my beloved Son in whom I, the Father, realize myself as Self." A tremendous step in evolution happened then, down on the Earth. The Heavens revealed themselves. They were present in the being of Christ as Self, as I. This is expressed, as it were, exclaimed, by this Saturn who carried the sadness of the cosmic meaning of the Baldur-Hoedur myth. Now he realizes, Heaven and Earth can be united again.

In connection with Cancer, we find in Norse mythology another wonderful story that can explain what happened. That constellation of Cancer is connected with the so-called Bifrost Bridge. This is the bridge that spans the abyss between Asgard, the dwelling of the gods, and Midgard, the dwelling of the human race. At the time of the Twilight of the Gods this bridge was burnt down, so the myth says. So there was no longer a bridge. What does this mean? The myth wants to tell us that the nature-given connection that existed in ancient times between the Spiritual World and the human world—so that the human being could look directly into the spiritual world—was interrupted. After a certain moment it no longer existed. But now, during the Three Years the bridge was rebuilt, the kingdom of heaven came again close to humanity. Heaven was open again to men of good will. And then at the time of Whitsun, Leo was given a new meaning. The original meaning was this: Once upon a time in the course of creation the periphery was gathered together in order to build the human heart—originally the archetypal functions of rhythm were right out in the periphery of the cosmos. You find corresponding indications in the book *Occult Science* by Rudolf Steiner. These functions eventually became enclosed in the chest, the "house" of the human being. From the Christ events on, they can again expand toward the periphery. The human being can grow again toward the cosmos, but now in such a fashion that the integrity as a self is maintained. All this was spoken into the movement of Saturn during the Three Years by the deeds of Christ.

Now we go on to Figure 1.2. We look at the same thing from the heliocentric point of view, that is, from the viewpoint of the Sun. Again we have here Cancer and Leo. During the latter

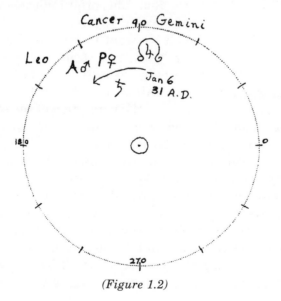

(*Figure 1.2*)

part of the Three Years, Saturn moved through the perihelion of Venus and eventually through the aphelion of Mars. What does this mean? This is one of the most important aspects of the story and can also give an idea of how one can combine, can make, the geocentric and the heliocentric cooperate. For our present purposes, there is no need to insist exclusively on one of the two. To maintain that the world is heliocentric and that is the end of it, or that it is geocentric and we cannot do with the heliocentric, is in our present context immaterial. We can make the two cooperate, and then we can get the most wonderful interpretations. What do we mean by this? The planets move, according to the heliocentric point of view, around the

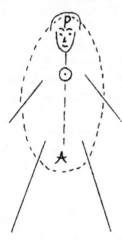

(Figure 1.3)

Sun. They do not move around the Sun in circles but in ellipses. The Sun is in one of the two foci (Fig.1.3). The ellipse is the orbit of the planet. It is overdone in the figure for the sake of demonstration. In some cases the differences are much smaller. The perihelion is the position where the planet is closest to the Sun whereas the aphelion is the position where it is furthest away from the Sun. This orbit indicates the sphere of the planet.

If we want to come to a deeper comprehension of cosmic events, we must enter an investigation of the spheres. The spheres are contained within the orbit of the planet. With this concept we come closer to the perception of the cosmos as a living organism. This endeavor is most important just in our time. We know that men have landed on the Moon, and there was much speculation on what this means for our knowledge of the cosmos, how this will affect the Moon, and so forth. However, we must not forget that the Moon is only a reflection. The Moon that we see in the sky is only the visible indicator of that whole invisible sphere that is contained within the orbit of the Moon. This sphere is the more important element. It is, so to speak, the workshop of Divine Hierarchies. This is where they work. Thus, for instance, the beings who are connected with

Venus have their workplace, so to speak, in the sphere of Venus. The beings of Mars work in the sphere of Mars, and so forth.

For our work it is important to know the difference between perihelion and aphelion. They are deeply connected with that which was introduced into the universe and evolution around the time of the Fall in Paradise. Before the Fall the cosmos was, so to speak, an integrated whole, and humanity lived in this wholeness. At the moment of the Fall humanity came apart. On the one hand the head forces developed. In other words, humanity "ate from the Tree of Knowledge." It is really a Tree of Knowledge, that which grows in the body. We need only imagine the central nervous system, the brain, as the root system of that tree and the nerves that spread like branches into the whole body. One thing was then withheld from humanity, the Tree of Life, which is the opposite, the other polarity of the human form that then developed. If we imagine the heart and the arteries going out from the heart, we have a picture of the Tree of Life, which spreads through the organism, as it were, against the Tree of Knowledge, the tree of the nerves. Actually the human form was thus distorted. The gods had created it differently.

This polarity was even introduced into the life of the spheres of the planets. Somehow all the spheres of the planets are involved in this. We recall again, for instance, the workings of Saturn, Saturn forming the skeleton. Here we can quite clearly get an impression of the polarity. On the one hand is the skull built by Saturn, which encloses the brain. And on the other hand, the skeleton of the limbs supports the muscular substance, and so forth, from within. This is connected with that stage of evolution around the time of the Fall. Thus we can say that at perihelion the planetary sphere is really integrated into our solar system. The spiritual beings who are connected with the perihelion within a particular sphere are interested in this solar system. They combine with it, identify themselves with it. In other words, we have something like the cosmic equivalent of a head (Fig. 1.3). It is an activity, somewhat faintly similar on a cosmic level with what every human being does with the head when observation combines with perception. In the opposite part, at aphelion, the planet and its sphere almost look as though they were intent on going off on a rampage in the greater cosmos. It expresses at least the tendency for this, whereas at perihelion, the Sun would hold it back within the solar system. In a certain sense, an element of stagnation is involved in this polarity; an element, however, that facilitates a certain evolution until such times in the future when a breaking out of this state of stagnation will be necessary.

Thus we see here something wonderful during the Three Years: On the one hand Saturn moves through the perihelion of Venus and on the other through the aphelion of Mars. The "head" part of the sphere of Venus and the "limb" part of Mars are engaged. In this is expressed something of tremendous import, even right into our time. In *The Changing Countenance of Cosmology* we spoke about Mars. In its movements, in the rhythms of its conjunctions and oppositions to the Sun, it forms two squares, one set upon the other (Fig. 1.4). This is the prin-

ciple of Mars, which means that it is really set and firmly fixed, consolidated, and absorbed into matter so that it is no longer capable of change and movement. Mars is continually doing this in the world and in us. With Mars is connected the sense of material existence. By following the Eight-fold Path of Buddha, one does actually

(Figure 1.4)

redeem the double square and make it pliable for seeing the reality of the material world. We need Mars. Every human being needs Mars. This planet is connected with aggressiveness. It is the god of war, whom we need in order to use our senses properly. If I look at certain objects, I must put these objects into their proper places, distances, and so on. I say, "You are there and I am here, and only on that level can I see you." Otherwise I am in danger of dissolving and melting together with the objects. And I would not see them as a self. So, we need Mars, but it can go too far and bind us forever to the material world. In other words, we would get stuck in the first part of the Earth's evolution. During the first half, the Cosmic Divine Wisdom was incarnated step by step, precipitated into matter. This is the First Creation, which is described in Genesis.

Venus, which we call the "Evening Star," was in very ancient times called "Mercury." The names were exchanged at some point far back in history, Rudolf Steiner says. In *The Changing Countenance* we worked out the elements of this planet that astronomy calls Venus. Thus we came to realize that the conjunctions of Venus with the Sun inscribe a pentagram into the ecliptic. This is a picture of the perfect integration, perfect harmony, within the human form, and still more, within humanity. In other words, it pronounces love and harmony in a cosmic sense. Traditional astrology also speaks of Venus being connected with love in a more worldly sense. This is correct, but it is only a tiny fraction of the reality of Venus. It is really that impulse of love and compassion, communication with all existing things, which was introduced and practiced by true Christianity.

Saturn moved through this cosmic expression, through the perihelion of the impulse of love with regard to all future Earth evolution. It is the redemption of Mars, which would want to go its own way of material, loveless, and cold factual existence only, expressed, for instance, in all that happens in its perihelion-aphelion line. Thus Saturn moved, about eight months after Golgotha, through the aphelion of Mars. These lines are moving through the zodiac in time. At the time of Christ, the perihelion of Venus was in Cancer, where it still is. However, at present it is further toward Leo. The aphelion of Mars, the faster one, was then already in Leo, but the two were then closer together than at present. Indeed, if we did go back into pre-Christian times, we would find that at one point in the history of humanity the two coincided. So, one would find the perihelion-line of Venus being identical with the aphelion-line of Mars. The two were, so to speak, at loggerheads. The one had the "head" part of its sphere where the "limb" part of the other sphere was located. There existed a contradiction between the two spheres, a fundamental contradiction in cosmic terms. How did this express itself?

When in ancient times the great initiates of humanity had problems, they put their questions to the cosmos. They asked, so to speak, the Sun through the medium of the Moon. We can no longer do this. We must, in order to receive answers, look toward the "connections in time"—no longer to space, but time. In other words, what we are going to do now in order to find an answer to our problem, the meaning of that coincidence between the spheres of Mars and of Venus, is to go back in time. We go back to that moment in history when the two actually did coincide. There we find something very interesting. We discover that they coincided about 3200 to 3100 B.C. This is the beginning of the Kali Yuga, the Dark Age. Thus, at the beginning of the Kali Yuga you have that unique event. It takes some of these "conjunctions" tens of thousands and hundreds of thousands of years to recur. Thus, we can see how the Christ Impulse "timed in" with the historical and physical consequences of the Dark Age of oncoming materialism, in order to lay the foundation for the slow redemption of Mars by the forces of Venus, or, in an occult sense, by "Mercury," the healing impulse of the cosmos. This we can read in the position and associations of Saturn during the Three Years.

2.

The Movement of Jupiter during the Three Years

IN THE LAST CHAPTER we spoke about Saturn. Today we shall go one step further, to Jupiter. What is Jupiter's work, or task in the human organism? It lays the foundation in the prenatal time for building up the whole nervous system. It also has some connection with the muscles. In our conscious realization it becomes the power of logicism, philosophically speaking. Saturn is connected with Gnosis, in a philosophical sense. Thus Jupiter can become the capacity of thinking. This is also associated with the future of our universe. On the present Jupiter already exists the place—one could even say the laboratory—for the preparation of "Future Jupiter." (See *Occult Science*.) There are beings dwelling on Jupiter who prepare that future and that future universe, which will come into being after the present one will have vanished. However, that future Jupiter will no longer consist of the kind of matter that we find here on the Earth. That will be gone and the future cosmos will be composed of thought. Rudolf Steiner once pointed out that just as we now find on the Earth geological layers that tell us of the past of the Earth, of past stages of evolution, so will we find on Future Jupiter "layers of thought." What we are thinking now, during the Earth evolution, will become the strata of that future universe. For instance, the materialism in which humanity is involved at present will be something like a hardcore thought layer on Future Jupiter. Thus the present planet, Jupiter, is connected with the future, whereas Saturn carries the memory of the past.

We are now going to work this out with regard to the hierarchies. The Hierarchy of the Seraphim, the Cherubim, and the Thrones (Fig. 2.1) is the Hierarchy of the Divine Father. Above them is the Trinity of Father, Son, and Holy Spirit. But the Father Forces would especially work with this hierarchy. Then follows the Hierarchy of the Kyriotetes, the Dynamis, and the Spirits of Form, or Exusiai. This is the Hierarchy of the Son. Closer to humanity stand the Archai, the Archangeloi, and the Angeloi, the hierarchy through which the Holy Spirit works. These hierarchies work into the spheres of the planets. Thus the Thrones work into the sphere of Saturn—the whole sphere, not only the planet. This concerns all the implications contained in the astronomical elements of the sphere, for instance, the lines of the nodes and perihelion-aphelion. The Kyriotetes work through the sphere of Jupiter, the Dynamis through Mars. The Exusiai have, so to speak, their workplace in the Sun sphere, the Archai in astronomical Mercury, the Archangeloi in astronomical Venus, and the Angeloi in the sphere of the Moon.

	Seraphim	
FATHER	Cherubim	
	Thrones	♄
	Kyriotetes	♃
SON	Dynamis	♂
	Exusiai	☉
	Archai	☿
HOLY SPIRIT	Archangeloi	♀
	Angeloi	☽

(Figure 2.1)

We are concerned at the moment with Saturn, about which we talked yesterday, and Jupiter. We see there the transition from the Father Forces to the Son, from the Thrones, or Spirits of Will, to the Kyriotetes, or Spirits of Wisdom. Of course, we do not identify the hierarchies with the spheres. The latter are the workplaces of the hierarchies. The hierarchies have, in a sense, their domiciles outside the solar system, in the sidereal world beyond the limits of our solar system. Beyond the solar system are the spheres of the fixed stars, and these are the dwelling places, so to speak, of the hierarchies. From there they work into the spheres of the planets, right down to the Earth. We can imagine how their hands, metaphorically speaking, reach down into the spheres of the planets and use them in order to bring about the developments that have to happen in our solar system. They remain with their being in the fixed-star world.

The offspring of the hierarchies are actually the beings of the elemental world. They are the descendants of the hierarchies. Of course, the good ones, who have not fallen in with the adversaries, assist the hierarchies in their work. For instance, the Moon is connected with the weather. This has now been officially established by scientific statistics. However, it is elemental beings, working from the Moon, who assist in this work of bringing about the weather conditions on the Earth, precipitation, and so forth. The decisions, the inauguration, and so forth, that is inspired by the hierarchies.

When the Cosmic Christ entered the corporeality of Jesus of Nazareth, He came down as a being of seven principles, similar to the human being who also carries seven principles, the lowest being the physical body. The highest principle of the human being is "Spirit Man." Likewise we can speak of a highest principle of the Christ in which He is the head of the Kyriotetes. Let us imagine that there are seven Kyriotetes—this is, in a sense, a generalization—who are working through evolution. They were already working on the Ancient Sun and at that time endowed the created world with life. The head of this whole hierarchy is the highest principle of the Christ. Therefore we can speak of the Kyriotetes especially as the Hierarchy of the Son.

We turn now to Figure 2.2. Jupiter moved during the Three Years from Aries into Gemini. It is connected with the workings of the Kyriotetes, the Spirits of Wisdom. Apart from this, Aries too is associated with the Kyriotetes. Here we are considering the fact, mentioned before, that the sidereal world is actually

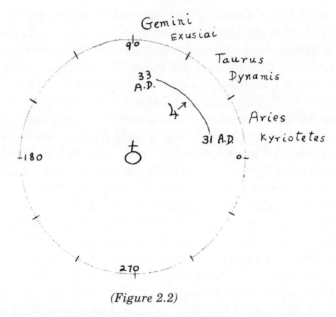

(Figure 2.2)

the world of the dwelling places of the hierarchies. Thus the constellations of the zodiac are in a sense the "houses" of the hierarchies. They step forth from these "houses" into the solar system and work there. In this sense Aries is associated with the Kyriotetes, Taurus with the Dynamis, and Gemini with the Exusiai. We do not, of course, identify the constellations of the zodiac with the hierarchies. They can only be considered as domiciles, or gates, through which the hierarchies work. Jupiter in Aries would then indicate that at the moment of the Baptism the head of the Kyriotetes descended into the body of Jesus. The time following the Baptism, actually the major part of the Three Years is devoted to the manifestation of Jupiter in Taurus. Taurus is connected with speech, with the larynx, and also with hearing. It is a manifestation of the Word, the Logos Who then spoke to humanity in the words of Christ, the curative and healing Word. Then, at a certain moment, Jupiter stepped into Gemini. Yesterday we found that Saturn started in Gemini. That was a different proposition. That was, so to speak, an expression of the "Great Reproach" of the divine cosmic world. Humanity had in the course of its descent into Earth matter torn asunder heaven and Earth, had lost the kingdoms of heaven. That was the "Cosmic Accusation," if one can express it like that, which spoke through Saturn, the representative, or workplace, of the Thrones, the Spirits of Will. But when Jupiter stepped into the constellation of Gemini, it meant something else. We must apply a different perspective. Gemini stands for the Two (Fig. 2.3.), the heavens above and the Earth below, from the viewpoint of Saturn. There exists, however, another "twin-proposition": a horizontal side by side, the principle of the "brother." This no longer falls in with the hierarchy principle. In the sense of hierarchical order, one was placed above the other—one was giving commandments to the lower one. The brother

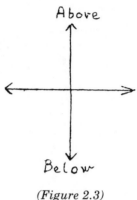

(Figure 2.3)

principle was made manifest during the Three Years by Christ at the moment of the washing of the feet. After that He said to the disciples: "Ye are my friends, if ye do whatsoever I command you. Henceforth I call you not servants...but I have called you friends." (John, 15, 14-15). In that moment Gemini-Jupiter was given a new meaning by Christ. The perspective of a new order, in the place of the old hierarchical order, was introduced by Christ. It is the beginning of a new cosmos which can be created if human beings will fall in with this deed.

Thus we are faced here with a new aspect. Yesterday we said that Saturn was at the moment of the Baptism (Jan. 6, 31) in the ascending node of Jupiter, in Gemini (Fig. 2.4), close to the transition from Gemini to Cancer. Jupiter stepped into its own node in 33 A.D. At the beginning, that is at the Baptism, the Father Forces let the Son descend with the words: "This is my beloved Son in whom I realize myself as Self." This is expressed in the heavens by the position of Saturn in the node of Jupiter. At the moment of Golgotha, Jupiter steps into that place. Now the Son is revealed, the Kyriotetes, He who had already enlivened the created beings on the Ancient Sun and who prepares the Future Jupiter. He is revealed in that moment, during the three days from Good Friday to Easter Sunday morning, the time of the Resurrection.

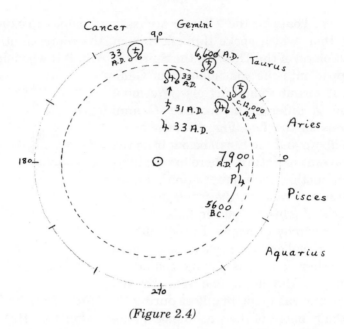

(Figure 2.4)

At this point it might be a good idea to study the history, or "biographies," of the two spheres involved here, those of Saturn and Jupiter. We might thus get a glimpse of that which worked then into the solar system from the heights of the hierarchies. We go now to the movements of the elements of the spheres, first that of the perihelion of Jupiter, which is in Pisces at present. It entered Pisces very long ago, in about 5600 B.C., coming from Aquarius. This transition was obviously connected with an important moment in the evolution of humankind. How can we gauge this point in history? In order to answer this we employ the precession of the vernal point. In 747 B.C., the foundation of Rome took place. The vernal point (position of the Sun on March 21–22) was then in the constellation of Aries. At present it is in the constellation of Pisces, and some time in the future it will move into Aquarius. In 747 B.C. it was in Aries; 2,160 years earlier it was in Taurus. That was the age of Egypt and Chaldea. Another 2,160 years back brings us to Ancient Persia. Now we have arrived at 5067 B.C. In order to move further back, to 5600 B.C., we must enter the preceding age. That is Ancient India. Thus the crossing over of

the perihelion of Jupiter happened 600 years prior to the commencement of Ancient Persia. Ancient India was a civilization
that was still closely connected with the awareness of the spiritual world. Many stories exist which confirm this. For
instance, the Holy Rishis then still spoke at certain times
directly out of the presence of the planetary genii. Thus they
stepped forth and taught the people. The Ancient Indian civilization was still very much aware of the reality of the spiritual
world. In such a world situation it happened, toward the end of
that civilization, that the perihelion of Jupiter moved into
Pisces, where it is now. It will move into Aries in 7900 A.D. This
brings us to the end of the Fifth Epoch:

1413 A.D.	Beginning of Piscean Age
2160 years	Present
2160 years	Aquarian Age
2160 years	Capricorn Age
7893 A.D.	End of Fifth Epoch—Post Atlantean

We start out from 1413 A.D., which is the commencement of
the Piscean Age. We add another 2,160 years. This brings us to
the Waterman Age. From there we move forward another 2,160
years. Now we have arrived at 7893 A.D. That will be the end of
the Fifth Epoch, the end of the so-called Post-Atlantean Epoch.
An epoch comprises seven Ages. The first Post-Atlantean Age
was Ancient India, followed by the Ancient Persian, Egyptian,
and Greco/Roman Ages. The Fifth Age is our own. Two more
are to come, the Age of Aquarius, followed by the Age of Capricorn. Almost exactly at the termination of the Fifth Epoch and
the commencement of the Sixth Epoch, the perihelion of Jupiter, the head part of that sphere, will move into Aries. What
could this mean? It is described in the Revelation of St. John.

Actually it is the beginning, the mighty opening, of that
whole Sixth Epoch of seven subdivisions that is described. The
writer of Revelation says: "And I saw, and behold a white horse:
and he that sat on him had a bow; and a crown was given unto
him: and he went forth conquering, and to conquer." (Rev 6:2)

It is called the opening of the first seal, and it refers to the moment when the vernal point will be in Sagittarius. Sagittarius is the man sitting on a horse who has a bow; that is how he is depicted on old star maps, half horse and half man, or the Centaur. But it is a white horse. In mythology the horse is always connected with intelligence, with the development of intelligence. In this case it is "white intelligence," in which we see the intelligence of Jupiter. This white intelligence will then also be associated with the perihelion of Jupiter, which will by that time have moved into Aries. Behind Aries, we said, stand the Kyriotetes, whose head is the Christ. All this points to that humanity that will then have accepted the Christ and will live in His Intelligence.

Now let us look at the nodes of Jupiter. The nodes are the crossing-points of the orbit of the planet through the plane of the movement of the Earth, the ecliptic plane. At present, the ascending node of Jupiter is where it was also at the time of Christ, in Gemini (Fig. 2.4). The ascending node of Saturn was then still in Cancer. Now it, too, is in Gemini. So the two are close together. They were then, at the time of Christ, only about 16 degrees apart. In the meantime they have moved even closer, and there will come a time in the future when the two nodal lines will actually join. This will happen, according to the present rate of movement, in c. 12000 A.D. We are here confronted with truly apocalyptic time conceptions. The figure for the end of the present Fifth Epoch is 7893 A.D.:

7893 A.D.	White Horse	(1st Seal) V.P. in ♐
2160		
10053		
2160	Red Horse	(2nd Seal) V.P. in ♏
12213	—Joining of ♄ + ♃	
2160	Black Horse	(3rd Seal) V.P. in ♎

The following 4,360 years brings us to the era of the Red Horse (Revelation 6: 2), lasting until 12213 A.D. This would be just the time when the two nodes are expected to coincide. So, we have to imagine a joining up of forces, of the spheres of Jupiter and Saturn, as far as the Earth is concerned. What will happen then? The Revelation of St. John says (Chapter 6): "And when he had opened the second seal, I heard the second beast say, Come and see. And there went out another horse that was red: and power was given to him that sat thereon to take peace from the earth, and that they should kill one another: and there was given unto him a great sword." We need also to know what it said about the third seal because the cosmic event of which we are speaking will happen just on the point of crossing over into the third seal from the second. Those seals comprise whole Ages of 2,160 years each. "And when he had opened the third seal I heard the third beast say, Come and see. And I beheld, and lo a black horse; and he that sat on him had a pair of balances in his hand. And I heard a voice in the midst of the four beasts say, A measure of wheat for a penny, and three measures of barley for a penny; and see thou hurt not the oil and the wine." The picture of the balances refers to the vernal point being then in Libra; the red horse is Scorpio, the white horse Sagittarius.

It is interesting to note what is said about the vision of the red horse. This is a moment when Judgment Day seems to have arrived. They will kill one another, those who have not followed the white horse, the White Intelligence of the Christ as it should be evolved by humanity. This is timewise associated with the coincidence of the two nodal lines of Saturn and Jupiter. Saturn is the great Judge, and from Jupiter streams forth all that which is connected with the Christ Impulse and the future of the Earth. The intelligence of the human race that will then not have become "white" will be "black." As the black horse it will then be bound to black and unredeemed earthly matter. And it will go down with matter. That will happen about 10,000 years from now.

Thus we see how the heavens reflected the events on Earth at the time of Christ. They proclaimed that the great Judgment Day in the far future was already preordained in its first stages by the Deed of Christ. He provided humanity with the opportunity to raise itself to higher planes of existence. Finally, He demonstrates it in the act of Resurrection. And yet, He left this humanity of all ages to come entirely free to decide on its own grounds. There was nothing of a compelling or imposing nature in the Deed of Christ. It is all up to our free decision and our love of the will of the divinity. Our thinking comprehension must be the first step.

However, the divinity cannot force salvation on us, nor can it preserve us from the consequences if we reject it. In the inevitable movement of the nodal lines of Jupiter and Saturn toward their coincidence is expressed the apocalyptic perspective of what will happen if humanity accepts or rejects the Christ Impulse. Those who accept will experience at some future point the most powerful combination of an awareness of the "Will of the Father" (Saturn) and a loving new "Creation by the Son," the New Adam of St. Paul (Jupiter). This will become part of their own being. The others who reject will experience that judgment of which the Revelation speaks, which will come in stages. Without the Christ Impulse, existence in any form will make less and less sense, until stages will be reached where these others will destroy themselves.

3.

The Movement of
Mars during the Three Years

WE HAVE WORKED during the last two sessions with the movements and associations of Saturn and Jupiter during the Three Years. We did this not just in order to have the information. It has a definite, practical relationship to our present age. In the course of this present century the Christ events, the Christ Impulse, have again come close to humanity. During the last nineteen hundred years humanity had to live on tradition. It had mainly only the documents, and we have experienced how this tradition is coming to an end. It is getting short of breath, in a sense. We see this happening all around us, in the ailing of the churches. It concerns nearly all religions, but the Christian religion seems to suffer most from this decline. And just in this situation, which has arisen in the course of this century, we can lift up our eyes again with hope. Rudolf Steiner revealed already in 1910 that during this present century will commence the experience of the Second Coming of Christ, the Presence, or the Parousia as it is called in the Greek version of the Gospels. Rudolf Steiner pointed out that this Second Coming will occur in a life or ether body. This would explain the corresponding expressions used in the Gospels: Luke, 21:27, "And then shall they see the Son of Man coming in a cloud with power and great glory"; or Acts 1:9, "A cloud received him out of their sight...[He] shall so come in like manner as ye have seen him go into heaven."

From the year 1935—actually from the years 1933, 1935, and 1937—onward into the future and during the next 3,000 years, humans in increasing numbers will be able to have the experience of the closeness of the Risen Christ in the ether body. Rudolf Steiner pronounced this on the basis of his spiritual investigations. The tremendous and dramatic pictures, words, and deeds that happened on the Earth during the Three Years are not lost in the cosmos. After some sojourn and development in the wide cosmos, where they were strengthened, they have now come back again to the Earth. They are in the neighborhood, or in the aura, of the Earth and can work down into the spheres where human beings live. Thus can the Christ manifest again as the Risen One, who is close to a humanity which seeks Him. We have experienced that these events of the Christ are, as a rule, accompanied by the ever-repeated configurations in the heavens similar to those that took place during the Three Years. Thus we can look at the rhythms and workings of Saturn and Jupiter during the Three Years as something like a foundation, as principles or archetypes, for all the future. These archetypes can be with us, and if we study Saturn, for instance, in our days and look particularly for returns to the original positions, then we shall find in history, and particularly in modern history, moments in which the Christ is especially near to a humanity that accepts Him. We shall demonstrate this.

As we followed the movements of Saturn and Jupiter during the Three Years, we came to the conclusion that during the final events the two planets were in the two constellations of Cancer and Gemini (Fig. 3.1). In the latter constellation we found the ascending node of Jupiter. The node of Saturn was still in Cancer. Furthermore, we found that Saturn moved between the perihelion of Venus and the aphelion of Mars. During the Three Years there is one more thing that is important, the node of Neptune. We must also include the spheres of those outermost planets that were discovered only during the eighteenth, nineteenth, and twentieth centuries, that is, Uranus, Neptune, and Pluto. They joined the solar system late, but we can assume that they were already present at the time of Christ. Saturn and Jupiter moved through the zodiacal region of these planetary elements during the Three Years. Both planets return to these positions periodically. Saturn gets there every thirty years and into the opposite point—we must

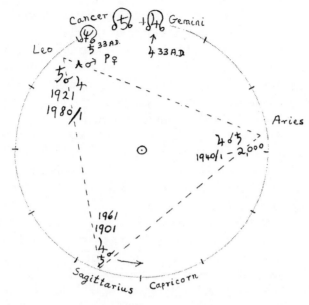

(Figure 3.1)

include them too—every fifteen years. Jupiter is faster. Every twelve years it would return to the original position. To this we add now one particular feature that is important for the life of the cosmos. These are the so-called Great Conjunctions. In the course of sixty years, they happen every twenty years in three points of the zodiac. This triangle rotates in time. Take, for instance, our present age. In 1901 there was such a Great Conjunction of Saturn and Jupiter in the constellation of Sagittarius. The two planets were then close to the descending node of Jupiter. Sixty years later, in 1961, the Great Conjunction happened again, and in 2020 it will repeat itself. But we discover that the latter two conjunctions have, or will have, moved deeper into Sagittarius, away from the descending node of Jupiter. The one in 1961 was in the descending node of Saturn, and in 2020 it will be close to the aphelion of Venus. So, there is movement, and therefore the whole triangle that is inserted in the zodiac by these conjunctions moves. It is like the hands of a clock. Thereby we can define and gauge time in a historic sense. This triangle needs about 2,500 to 2,600 years to go around once through all the constellations of the zodiac. Thus it happens that at certain times these Great Conjunctions are either moving through the places of the ascending or descending nodes of Jupiter and Saturn, and through the perihelion or aphelion of Venus. In 1901 when the Great Conjunction took place in the descending node of Jupiter, such a cycle of events began. In such a moment the heavens are, so to speak, open to an activation of Earth events by a cosmic commemoration of the Christ Event. This commemoration was very active in history around 1901. In 1899 the five thousand years of Kali Yuga, the Dark Age, had come to an end. (At least it should have come to an end. Sometimes one can have the impression that we have dropped even deeper into darkness since.) In 1900–1901 Rudolf Steiner started to communicate his message of Anthroposophy to humanity. At that time he laid the foundation for a spiritual understanding of Christianity specifically by giving a cycle of lectures in Berlin which were then collected in the book *Christianity as Mystical Fact.* Thus do these cosmic commemorations work in history and become reality on Earth. In this way we can indeed speak of a return, of a presence, of the Christ in some form. So we can go on to 1961 and ask what it might have meant for humanity. A very objective observer could discern a continuation of that which was started in 1901, and would detect that at the end of 1930 a heliocentric great opposition had taken place, with Saturn near the descending, and Jupiter near the ascending, node of Jupiter. (These are "halfway houses" in the cycle of sixty years.) One would then discover that humanity around that time began a kind of move toward a World Golgotha. We need only think of the events in Central Europe during the thirties. During the years since 1961 this move seems to have accelerated markedly. And yet, in all the gravity of external developments we can all the more hope that a World Good Friday will be followed by a World Easter Sunday, by a revelation of the power of Resurrection.

If we go through history we find similar occasions earlier. Already during the first centuries after the Mystery of Golgotha, about 134 A.D., and then at intervals of sixty years (that would be 193, 253, 312, and 372), Great Conjunctions occurred. They took place in the opposite part of the zodiac to where the two planets were during the Three Years. In the opposition they worked all the more as activating commemorations but then rather in individuals. The conjunction in 312 A.D. is associated with the Emperor Constantine. It was the moment when he was converted to Christianity. The era of the persecutions of the Christians came to an end, and Christianity became a state religion. Surely, thereby, a kind of Christianity was inaugurated that was not always conducted in the spirit of the original, esoteric Christ Impulse. But behind this development stood at least faint experiences of the latter's reality.

Later, starting in the sixth century, the Great Conjunctions of Saturn and Jupiter occurred in the ascending nodes of Saturn and Jupiter, also in the other planetary elements that we mentioned. These were the places in the zodiac where the two planets actually moved during the Three Years. In 590, and after that at intervals of sixty years, the Great Conjunction arrived in these positions, in 650, 709, and 769 A.D., right into the ninth century, in 829 and 888. These were very important moments in the history and development of Christianity. Around 590, actually in 597, Augustine of Canterbury arrived on the island of Thanet, in the estuary of the Thames, from where he started to convert the population of the British Isles to Roman Christianity. This was essentially founded on the tradition of the other Augustine, Augustine of Hippo, who taught on the basis of the realization that all direct insight into the Christ Events, which the first Christians had, had been lost. All that was left, in his mind, were the Scriptures. This was the beginning of dogmatic Christianity, and the Roman Church increasingly evolved its existence and its work on this attitude. The second, later Augustine founded Canterbury. Soon after that a war of extinction was started against the Celtic Church in the north. It lasted many years, but eventually the Celtic Church, which had come from Ireland and had its center at Iona, was extinguished. It was a kind of Golgotha event for esoteric Christianity, because Celtic Christianity was an esoteric Christianity. It was able to live in that Presence of which we spoke earlier. Many stories exist that seek to describe this fact. St. Bridget of Iona is one such example. The story relates that she is supposed to have traveled in a spiritual form to Palestine, at the time of the birth of Jesus. She relieved the mother of Jesus for one night, and in the morning she received from Mary her mantle, in thanks. Into this mantle were woven stars, which obviously emphasizes the cosmic background of the birth of Jesus. Then there is the story of a king of Ireland who had a visitor. He went with that visitor for a walk in the woods. Suddenly the King rushed up against trees, hitting about him as if he had gone mad. When he was asked, he said that in that moment the most atrocious deed in humanity on this planet had been performed. It was the moment of the Crucifixion, and that event had enraged him fearfully. These stories attempt to tell us that the Celts were able to follow the events in Palestine clairvoyantly. On this was built Celtic Christianity, the Christianity of St. Columba and the other Celtic Saints. This was extinguished externally. It was a kind of Golgotha Event, followed by a Resurrection Event. Such examples in history can give us confidence that however dark a certain age might be, there can be hope for the future.

Rudolf Steiner spoke, on the basis of his spiritual investigations, about the elimination of Celtic Christianity. He pointed out that in this moment the Celtic Folk Spirit performed a sacrifice and renounced its task as Folk Spirit to become the guiding spirit of esoteric Christianity. This is evident from a careful observation of the corresponding historic events. In 709, one of the Great Conjunctions of this series occurred. Soon after, in 711, the Islamic Moors from North Africa broke into Europe via Gibraltar. They conquered the whole of Spain and proceeded to invade the realm of present-day France. About that time the foundation of a true antidote against the spreading militant Muhammadanism, the Order of the Holy Grail, was established. Charibert of Laon, the grandfather of Charlemagne, travelled to the East. And from the East he brought the message of the Holy Grail and all that it concerned. In the Holy Grail is indeed present the essence of esoteric Christianity, in which now the Celtic Folk Spirit could dwell and work. This is the background of the stories of King Arthur and the Parzival events. The latter took place during the ninth century, and the Great Conjunctions of 829 and 888 seem to be connected with it. They were indeed activating commemorations in the cosmos of the Resurrection and Whitsun events.

Another set of Great Conjunctions, carrying a similar power of commemoration, occurred during the eleventh and twelfth and right into the thirteenth centuries. In Christianity there raged then great disputes on the matter of the Transubstantiation. For instance, a personality like Berengarius was involved in these disputes. He was unable to conceive of the Transubstantiation of bread and wine into the Body and Blood of Christ at the ritual of the Holy Mass. Then we come to the fifteenth century, and there we again find Great Conjunctions appearing close to the nodes of Saturn and Jupiter, and to the perihelion of Venus and the aphelion of Mars. The node of Neptune is also involved. The dates of these conjunctions are 1444, 1504, 1563, and 1623. This was a time in history when esoteric Christianity, which had been reestablished by the movement of the Holy Grail, took another step forward. One can find a definite, historic sequence in the establishment of esoteric Christianity. The Holy Grail is a vessel that had been made, according to tradition, out of a jewel that fell from the crown of Lucifer in the heavens. It fell to the Earth and out of this was formed the dish that was used at the time of the Last Supper. It was then kept hidden for a long time, and its destiny was told by Robert de Boron, one of the French troubadours. According to the story, this vessel actually stems from the domain of the adversary Lucifer, who opposed the Christ during the Temptation and tried to divert Him from His deed of salvation. However, just this very substance of Lucifer is taken and transformed into the carrier of the holiest that Christianity possesses. The story says that every Good Friday a dove descends from the heavens and puts a Holy Host into that Vessel of the Holy Grail. It is the ever-feeding nourishment of the Grail knights. We find a transmutation of this principle in the Order of the Knights Templar. They founded the Order on the Holy Sepulchre in Jerusalem. It is a motif similar to that of the Holy Grail. The grave of the Earth, of the dark Earth, out of which the Resurrection took place, is made the foundation. It is a kind of transformed Grail motif. The Order of the Knights Templar was founded in 1118–9. It was destroyed, chiefly through the insistence of King Philip le Bel of France, at the beginning of the fourteenth century, finally in 1314. These dates are close to Great Conjunctions and oppositions near the descending node of Saturn, and between the descending node of Neptune and the perihelion of Mars. Again we see here a cosmic commemoration of the Three Years.

A further transformation of Grail Christianity took place in the development of Rosicrucian esoteric Christianity. In 1444 the Great Conjunction occurred in the ascending node of Jupiter. Fifteen years later Saturn was opposite, in the descending node of Jupiter. This brings us to 1459. These dates are intimately connected with the development of the Rosicrucian impulse. 1459 is the date that is connected with *The Chymical Wedding of Christian Rosenkreutz, Anno 1459*, which was published early in the seventeenth century. It reveals several profound aspects of a Christian occultism. In one sense it is the story of a great initiation into esoteric Christianity. Above all it confirms that Rosicrucianism is in perfect keeping with the principles of Grail Christianity and the principles of the Knights Templar. The black cross (of the Rose Cross) stands as a symbol for the dark Earth, for death, for all that we inherit from the material world. This is redeemed and sanctified by the wreath of seven glowingly red roses, the symbol of eternal life, of resurrection. Also in this imagination we can conceive of the transformation of the vessel by carrying the most holy spirit essence, which humanity received through the Christ Event.

Eventually we come to modern times, to the Great Conjunctions that we mentioned in the beginning. Again esoteric Christianity takes a new step forward. What Rudolf Steiner offered to modern humanity as Anthroposophy was not just a repetition or imitation of the Rosicrucian impulse, but a further development of Rosicrucian Christianity congenial to modern stages of human consciousness. Indeed, we can see that a study of the cosmic implications as we try to do it with regard to Saturn and Jupiter leads us to historic characterizations of the advance of the Christ Impulse in humanity. It leads us also to a comprehension of the great trials, the great Golgotha experiences, through which a humanity on this road must by necessity go, for the sake of its catharsis. In 1623 a Great Conjunction took place, close to the perihelion of Venus and the ascending node of Neptune. This happened at the beginning of the Thirty Years War. The position of the great opposition in

1930 in the history of modern humanity has been mentioned earlier.

We will carry on with Mars, by studying what Mars was doing during the Three Years. The very concept "Mars" implies the warlike attitude of this planet, aggression, and confrontation. Human beings need this, apart from the other gifts from the cosmos, in order to meet the Earth when they are incarnated. When human beings look at objects they, mostly unconsciously, push that object back into perspective and into a certain distance. Thus they maintain their integrity as individual selves. Through this process they are able to say: "There is the object, here am I." Otherwise, they might amalgamate with the object to such an extent that they could not recognize its independent existence. We need Mars in order to come to a clear perception of the physical-material world and to formulate speech. Mars is concerned with the development of speech, which is the highest manifestation of the forces of this planet in humanity. Thus there is always an attitude of confrontation connected with this cosmic heritage. For instance, the gall bladder is also associated with Mars. This organ does something similar in the digestive process. Foreign substances enter the body, for instance, as food. Mars then works through the gall bladder to confront, to attack what entered the body, at first as uncongenial substance. Otherwise the latter might poison the human organism.

This attitude of Mars can go, of course, to the extreme and threaten damage. This has happened in the course of history. Mars was deeply connected with the development of perception in the human race and therefore with the first creation, that is, the creation of that world which we now have around us as nature, including our own physical nature. But that first creation is now superseded by the impact of the Christ Event. St. Paul recognized this when he spoke of the second creation, of the Second Adam. In the course of the first creation, humanity entered ever deeper through sense perception into material existence. This was a necessary development in order to lead humans to independence. However, if this were to carry on beyond what is necessary, it would submerge them completely in matter. The Second Adam-man follows the Venus principle, the realization of love and compassion. As a teaching it was already prepared in humanity by the Buddha. For instance, the Eightfold Path of Buddha is one such means of preparation and redemption of a Mars that might have gone too far and become detrimental. This is the danger of the Mars capacities, that if they persist for too long they can become forces of obstruction, of opposition, against evolution. They want then to maintain their own existence. Therefore they may develop opposition to all that promotes the "New Spiritual Creation," the New Adam.

On the 6th of January, A.D. 31, Mars had just entered Aries (Figure 3.2). From there it moved around once through the zodiac and came back to Aries, where it went into a loop. That happened in the autumn of 32 A.D. Then it moved on into Gemini where it was on April 3 (at Golgotha) in 33. Thus right in the beginning Mars was in Aries. What does that mean? Mars tears down, as it were, the forces of Aries, leads them down, deep down even into the material world. Aries is in the cosmos

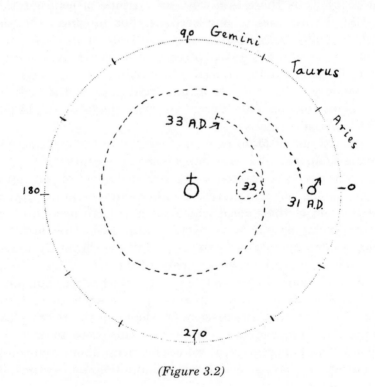

(Figure 3.2)

the archetypal region of the human head. The ancient symbol that we use is (Υ). It corresponds exactly with that which is inscribed in the physiognomy of the head. The brain rests horizontally on the base of the skull, above the eyebrows. From there a vertical line works down into the lower part, into the movable part, of the head. This is present in the downward trend, starting along the nose. If Mars works into this too strongly and too long, the cosmic image of the head may be adulterated. The emphasis may then be too strongly on sense perception, and particularly on reliance on the brain exclusively.

Mars was in this position around the time of the Temptation. The Christ was then confronted with the ancient heritage contained in the body of Jesus. This body of Jesus, like the "corruptible" body of any other human being, carried the involvement of being born into the realm of sense perception, into the material-physical world. It had to participate, by nature, in the consequences of the Fall in Paradise. The Christ faced during the forty days in the desert that which Lucifer had inflicted upon humankind in Paradise. He insisted that this material-physical world was his domain and that it must not be disturbed. Therefore he tempted the Christ, tried to lure Him away from his intended mission on the Earth, of bringing redemption even into this sphere of humanity's relationship to the physical-material world. Every human being is exposed to this temptation; mostly we don't notice it and fall. This is just a fact. By the Deed of Christ, by His rejection of the tempters, He laid the foundation for our return to our cosmic image, to our preordained position in a cosmic evolution. We had lost that and instead had to live in a corruptible body. In that moment of rejection of the tempters, a seed was, so to speak, lowered into the Earth, an example was set that was to prove that redemption is possible for every human being who would "follow" Him. Thereby was that Mars in Aries given the potential for a new meaning, which can be enacted in the course of individual human endeavor.

After that we see Mars move through the zodiac. It came back into Aries in autumn, 32 A.D., around the time of the Feast of Tabernacles, which is described in John 6, and performed a loop. Another test, similar to that which we described, took place. We read that Christ's "brethren" had gone up to the Feast of Tabernacles at Jerusalem. He stayed, at first, behind. But then we hear that He also went up unto the feast, "not openly, but as it were in secret" (John 7:10). He went in an invisible form into which He could draw Earth matter at His will to make it visible. This is the first glimpse of the "resurrection body," or the "phantom body," which the human being had lost through the Fall in Paradise. By overcoming the tempters earlier, He had taken a further step toward the reestablishment of humanity's cosmic dignity and potential.

We hear, however, in this Chapter 7 of St. John also of the growing fierce opposition by the old Mars-Aries forces. He says to the Jews (Verse 19): "Why go ye about to kill me?" It was the voice of the adversaries whom the Christ had overcome who spoke through the Jews. With all their might they wanted to hang onto the old "Mars" order and keep humanity in the bondage of perishable material existence. Finally we see Mars at the time of Golgotha. Here the constellation Gemini appeared, through Mars, as an image of the two adversaries, Lucifer and Ahriman. Lucifer had tempted Adam and Eve in Paradise, and he had succeeded. Thus also Ahriman found the door open. He had introduced death, and the fear of death, into humanity. Now, at the moment of Golgotha he again stood there, demanding his toll, but by the event of the Resurrection he was defeated. The "old death" had lost his power. Thereby the gate of hope was thrown open for all humans of good will, so that Ahriman and his impact on humanity can be overcome. Thus also was Mars in Gemini potentially redeemed.

If we could go into the details of the career of Mars during the Three Years we would find many more such instances of the redemption and salvation that were performed as potential by the Christ.

4.

The Sun during the Three Years

TODAY WE SHALL SPEAK about the Sun in connection with the Three Years. The Three Years must be regarded altogether as a kind of embryonic development. The Baptism was like a conception and Golgotha like a birth. It was the birth of the Christ Impulse on the Earth and in humanity. This "embryonic" development had the aim to establish in the end the resurrection body, the incorruptible body. There are many suggestions and much information contained on this in Rudolf Steiner's lecture cycle *From Jesus to Christ,* particularly in the latter chapters where he describes the nature of the incorruptible body. In the most ancient times, before the Fall, the human being had lived in such a body. But it was spoiled in the loss of Paradise. This spiritual form body was torn down into the realm where the laws of nature rule the world of physical matter. Through the Deed of Christ the seed was laid for times to come when it will be possible again for us to attain this incorruptible body that is not subject to the laws of matter, which means death and decomposition.

The moment of Golgotha, that is, the time from the 3rd to the 5th of April 33 A.D., displays a unique situation in the heavens. During those days there existed a harmony throughout the heavens, right down to the Earth, which did not exist before and will not come back to the universe for a long time, if it will recur at all. A period of about 26,000 years is involved in this rhythm, which is the precession of the vernal point. During such a time interval the whole composition and organization of the universe will fundamentally change, so we cannot be sure whether a similar configuration will return in the heavens. There was a harmony between the three zodiacs which we can discern in the universe. First there is the zodiac of the constellations, of the visible constellations which are indicated, drawn up, so to speak, on the map of the heavens by the positions of the fixed stars. Thus we distinguish the twelve constellations of the zodiac (Fig. 4.1). We start with Aries, Taurus, carry on with Gemini, Cancer, Leo, Virgo, Libra, Scorpio, Sagittarius, Capricorn, Aquarius, until we arrive at Pisces. Apart from this we recognize a second zodiac. This one is mostly used in traditional astrology. It is the zodiac of the "signs" of the ecliptic, which are employed in the newspapers almost exclusively. How does this come about? We imagine the path of the Sun in the course of one year. It completes a circle around the Earth,

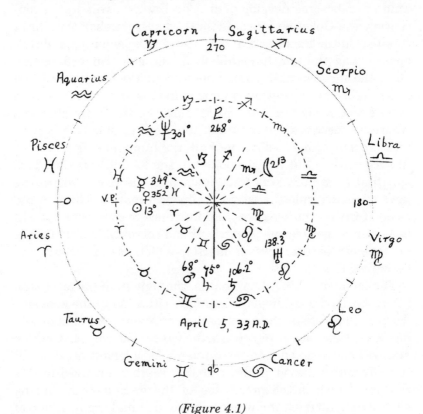

(Figure 4.1)

apparently, as we say according to Copernican astronomy. On the 21st of March of every year, the commencement of spring, the Sun arrives at a definite point in this circle. From this point we divide the latter circle into twelve equal parts, starting from that point where the Sun appears to stand on the 21st of March. So, we have here a circle that is closer to the Earth than the fixed-star constellations. The latter are deep in space. The astronomers tell us that some of the fixed stars are thousands of "light years" away from our solar system.

We have divided this inner circle of the ecliptic into twelve equal parts of 30 degrees each. These are the ecliptic "signs." One is almost inclined to say, it is unfortunate that they bear the same symbols and the same names as the constellations. Thus we speak of the sign of Aries, sign of Taurus, sign of Gemini, and so forth. Of course, this circle is not fixed forever. There is nothing within this universe in which we live that is fixed forever. Even the fixed stars move, and over thousands of years they completely change their relative positions within the constellations to which they belong at a certain time. Also this inner circle of the ecliptic changes in relationship to the constellations. This is brought about by the rotation of the axis of the Earth. The axis of the Earth points at present to the so-called pole star in the Northern Hemisphere. After some thousands of years one would notice that it pointed to another star, no longer to that which we now call the pole star in the heavens. So it pointed in previous millennia to different stars in the heavens. Thereby the position of the ecliptic is changed.

The ecliptic is, as we said, the (apparent) path of the Sun in one year. This we combine with the equator of the Earth, which we project into space around our planet. The two circles cross in two points. One point is the vernal equinox, that is, the point where the Sun appears to stand on the 21st of March. The other one is where the sun appears to stand on the 21st of September, the autumnal equinox. They are called equinoxes because on these two dates we have equal length of day and night. The vernal equinox marks the commencement of spring in the Northern Hemisphere. In the Southern Hemisphere this is reversed.

In the present age, if we could observe the Sun on the 21st of March and at the same time see the fixed stars far out behind it in space, we would make a remarkable discovery. We would not find behind this spring Sun the fixed stars of Aries. Although we know it is then in the ecliptic sign of Aries, it is not in the zodiac constellation of Aries. The stars of Pisces can be discerned behind that spring Sun. One would even find the right-hand or western end of that constellation behind the Sun. In fact, in a few hundred years we shall see the Sun on the 21st of March even still further back, right in the constellation of Aquarius, and so forth. This is the so-called precession of the vernal point. We use it as a foundation for the discernment of the Ages, of the civilizations of humanity. In Rudolf Steiner's lectures and books we read so often the expression: we are living now in the Age of Pisces, and the Age of Aquarius is to come. The Age of Pisces was preceded by the Age of Aries, and so forth. The precession of the vernal equinox is the foundation of all that.

Around the time of Christ, it was indeed thus that when the Sun on the 21st of March entered the ecliptic sign of Aries, one would also have seen behind the Sun, far out in space, the first stars of the constellation of Aries already coming up. So, signs and constellations coincided. There was a harmony, established once, within the course of about 26,000 years. We also have to take into account a third zodiac. This is the Earth zodiac, which is established by the rotation of the Earth. Wherever we are on the Earth we realize that the Sun has a certain relationship to our horizon. In the daytime it has reached a definite height in space. At the same time we would find a certain constellation rising on the eastern horizon, another one setting in the west, a third one culminating in the meridian of the place. From all this we could deduce that below the horizon the opposite constellation is deep down in the nadir of the place. The Sun does not stand still. It keeps rising higher till noon, when we find the Sun at the meridian. After that, in the afternoon it slowly descends toward the west, until it sets in the evening. Thus we have a set of space coordinates to deal with. Any place on the Earth has a meridian, a circle that rises from the south point, goes through our zenith directly above us, and then descends to the north point. From there it continues as a circle underneath the horizon until it comes up again at the south point. Our horizon constitutes another space coordinate. Thus the circle, which is the meridian of any given place, south point, zenith, north point, and nadir directly under us, together with the horizon, divides our regional geographic space into four segments. One segment lies between the eastern horizon and the meridian, another one between the latter and the western horizon. A third sector is situated between the western horizon and the nadir. This is followed by a fourth sector between the nadir and the eastern horizon. Thus we now have four segments, which look like the pieces of an apple cut into four equal parts. We divide all four pieces again into three subdivisions each, so that in the end the space around us looks like an orange divided into twelve segments. The axis of our space-orange would have to lie on the south-north axis of our particular geographical locality. These twelve subsectors are the so-called astrological houses.

This zodiac of "houses" radiates, as it were, from where we stand into cosmic space like the segments of an orange. The great configurations of the zodiac, the constellations, also the ecliptic and the planets, are received or organized into this space-organism. The "houses" of the third zodiac are like twelve receptacles that relate cosmic facts to Earth reality.

What do the three different zodiacs mean? The visible constellations of the fixed stars are the astral world. One can even say, with certain reservations, that this fixed-star zodiac is an image of the astral body of the cosmos. World astrality, world soul forces, move the stars. It is in this sense that we take the zodiac of the constellations as an expression of the astral body of the cosmos. The signs of the ecliptic rest on the vernal point in which the Sun indicates in the heavens the commencement of spring. This is a zodiac associated with the seasons, with the yearly rhythms of life in nature. It is particularly obvious in the plant world, as a kind of image of the ether, or life forces, in the universe. Here too we can, with certain reservations, say that this zodiac of the ecliptic signs is an image of the ether body of the cosmos, particularly of the solar cosmos. Finally, that which radiates as space segments from the geographical point where I stand is an image of the physical body, which is given us as earthly physical reality. They are also given the symbols and names of the constellations. The sector immediately below the

eastern horizon, the first "house," is called Aries, the following
one, deeper down, Taurus, the third Gemini, next Cancer, then
Leo, Virgo, and so on. The sector, or "house," immediately above
the western horizon is Libra, the following one is Scorpio. The
one that adjoins the meridian from the western side is called
Sagittarius. Then, toward the east follow Capricorn, Waterman,
and Pisces or Fishes. At the time of the Mystery of Golgotha,
particularly on Easter Sunday morning, before sunrise, these
space sectors were oriented into space in perfect harmony. Rising in the east were at that moment in the Earth-space sector of
Aries, the first "house," the ecliptic sign of Aries and the fixed
star constellation of Aries. So, the "astral" Aries and the
"etheric" Aries were received into the "house" or space-sector
Aries. With slight differences that one can calculate, all three
zodiacs were in perfect harmony. It will take almost 26,000
years until the same harmony can happen again in the cosmos.
But by then the conditions and the constellations might have
changed completely. They certainly will no longer resemble the
present shapes. In this sense a unique situation in this universe occurred at that moment when the Resurrection took
place. We remember from the Gospels that on Easter Sunday
morning two of the disciples went to the sepulchre and found it
empty, to their great amazement. The Resurrection had taken
place. This wonderful harmony of the three zodiacs, between
heaven and Earth, is a vivid symbol, or image, of the resurrection body that was then reestablished. The disharmony before
was connected with the "Fall of Man," and with the corruption
of the spirit body.

The Fall in Paradise did not mean that only Adam and Eve
went through those experiences of which we read in Genesis.
The whole universe was affected and changed in that moment.
The position of the Earth was changed. We are faced in our
work with the difference between the geocentric and heliocentric world conceptions. Rudolf Steiner, on the basis of his spiritual investigations, pointed out that in the medieval
Rosicrucian schools the pupil was first introduced to the geocentric view. Then, when the pupil had learned this, he was
told that this is how it should be, but that it is not so on account
of the Fall, on account of the "Sin of the World." The Earth had

thereby lost its central position. That which happened in Paradise affected the whole Earth and therefore the disharmony, the possibility of disharmony, entered cosmic history. However, on that Sunday morning just before sunrise, the original harmony was reestablished. It was, of course, not the cosmos that reestablished it, but the cosmos was jubilant about what happened when the Christ rose from the grave of the Earth. At the same time it was the hope of all future Earth existence and of all future mankind.

The distinguishing features of this event are the special aspects and the positions of the planets in the zodiac during those days. They were unique because the same things could not happen again for a long time. The Sun was at that moment in the house or space sector of Aries, in the sign of Aries, and in the constellation of Aries. What did this pronounce? It described what happened on Earth as a new beginning. The constellation of Aries is connected with the head. So this "head" or seed point in the cosmos was shining with the glory of that morning Sun. Next we find Mars and Jupiter related in that moment to the space sector Gemini, sign of Gemini, and constellation of Gemini. This is no longer the hierarchically oriented, vertical Gemini because the Washing of the Feet had taken place shortly before. It can now become a manifestation of the "brotherhood of man." This is one aspect, but there are, of course, more. We find another viewpoint if we go back in time with the precession of the vernal point. For instance, 2,000 or 3,000 years earlier we would find it in Taurus. That was associated with the Egypto-Chaldean Civilization. If we go still further back we discover the vernal point in the constellation of Gemini. That was the civilization of Ancient Persia. The leading ethical and moral motif of Ancient Persia was the world contradiction of light and darkness. The principle of light was the divine being, Ahura Mazdao, in the heights. His dwelling was not the Sun that we see, but rather the Ahura, the "aura," of the Sun, which was clairvoyantly experienced by Zarathustra in the heights of the cosmos. We saw that the Spirit of the Universe prepared for His incarnation as the Christ. And deep down in the Earth, He perceived Ahriman, the Lord of Darkness, who fights against the light. These are very practical views. The foundation of organized agriculture was built on them. By ploughing the soil we open it up to the light from the cosmos. We make the growth of the plant thereby possible. The human being's head is oriented toward the cosmos, but the plant lowers its head, the root, into the soil and stretches its limbs, that is, its trunk and branches, toward the Sun. The humanity of the future will learn to relate to the cosmos in full consciousness of self. This will be the Spiritual-Cosmic Communion of humanity. Rudolf Steiner gave some indications of this future perspective. Not for much longer will humans be able to passively watch the things that are going on in the heavens and ask: What will they do to me? Many still look up to the heavens with great apprehension. This is an attitude that we must overcome. The Christ Event can give us the ways and means to take up a positive relationship of spiritual activity toward the cosmos. "In times past the stars spoke to Man." They have become silent, at least to a very high degree. Now, and ever more in the future, "Man must learn to speak to the stars." This is not meant in a metaphorical sense. There is great reality in this perspective.

The two planets Mars and Jupiter were, as we said, in Gemini at Golgotha. Mars stands for the old or first creation, which must be superseded. Jupiter is a seed point for the future. In its sphere spiritual beings are already preparing Future Jupiter. All this is associated with the house, sign, and constellation of Gemini. From a certain perspective, Gemini can be connected with the hands. Painters in the Middle Ages who painted the Christ always presented this in wonderful gestures of His blessing, healing, guarding hands. Rudolf Steiner gave it expression in a huge statue that he carved. It shows Christ holding down Ahriman, who wants to chain humanity eternally to earthly matter. Above his head He holds back Lucifer, who would tempt humanity to forsake the Earth. At Golgotha Gemini was transformed into a threefoldness. Christ, the Savior of the human soul, created a balance of protection between the Twins of the Adversaries, Lucifer and Ahriman.

Venus and Mercury at Golgotha were on the point of rising, in the house, sign, and constellation of Pisces. In a certain sense, Pisces from any of the three perspectives reflects the end of an evolution. It is the end of the zodiac and also a new beginning. Of the two Fishes in this constellation, one swims in the heavens toward Aries, the other toward Waterman. They are connected by a faint ribbon of stars. Thus the past appears connected with the future. This is, for instance, the task of the present Age of the Fishes. We still have to overcome the past, of which much remains from the last Ages of Aries, and even of Taurus. On the other hand we must prepare the Age of Waterman-Aquarius, which is yet to come. Venus and Mercury, that is, the planet of spiritual love and compassion and that of the power of spiritual fire, had just risen on Easter Sunday morning. They were associated, so to speak, with the feet by being positioned in Pisces. They proclaimed in the heavens that the Risen Christ stood and walked on these principles. What can we read in this?

Gemini has been connected since ancient times with a vertical or hierarchical structure of social order. We need only think of the great empires in Ancient Asia where the potentate, or ruler, sat on his throne high above the common people. Many steps led up to it, on which the whole court was arranged according to rank. This was an image of hierarchy on Earth, an imitation of the divine hierarchies. This is contrasted by Pisces. There the two Fishes swim side by side. The two feet and the two hands of the human being, which are also an image of the two Fishes, are equal, though their functions may be distinct. As far as the social order is concerned, Pisces displays much more a principle of brotherhood.

An additional viewpoint is provided by the angular relationship between Venus in Pisces and Mars in Gemini. They were 76 degrees apart on that April 5th, 33. This is close to the pentagram or Quintile aspect. The pentagram-Quintile (Fig. 4.2) divides the ecliptic into five times 72 degrees. It is well known to astrology but not much used. This aspect is made by the conjunction rhythm of Venus, the planet of spiritual love and compassion. It confirms additionally that Mars received here an

(Figure 4.2)

impulse from Venus, an impulse of healing and progress. In one sense it is the overcoming of vertical social order by horizontal order, or brotherhood order. Saturn emphasized this even more by its position in Cancer constellation, Cancer sign, and Cancer house. Saturn, so to speak, proclaims in that moment that the spiritual power of the Christ Event can put the earthly house of humanity in order. Cancer is an image of this "earthly house." It manifests in the formation of the human chest, which is made of twelve pairs of ribs. If we imagine that this introverted region would open up through a powerful spiritual effort, it would appear like a Sun radiating out into space with twenty-four rays. Thus the narrow house of the human being would become a cosmopolitan house. This is one of the many imaginations that are latent in this Saturn.

5.

The Movement of Venus during the Three Years

TODAY WE SHALL LOOK at Venus during the Three Years (Fig. 5.1). On the 6th of January, 31 A.D., Venus was in Aquarius. From there it moved through the ecliptic and made a loop in Leo, around August, 31 A.D. After that it came into superior conjunction, which means it was behind the Sun, out in space. This happened in June, 32 A.D. At the time of Golgotha, just a few weeks before, it made a loop close to the vernal point, just between the constellations of Aries and Pisces. We shall insert two more features that were actually outside the Three Years. Before the 6th of January, 31, a superior conjunction took place in the constellation Libra close to Scorpio. Venus was then again beyond the Sun. That was on October 22, 30 A.D. After 33 a superior conjunction occurred in Capricorn, on January 8, 34. Both events were outside the actual Three Years but are still connected in a deep sense with the Mystery of Golgotha. Thus we get a definite and most important pattern. We simply mark the places in the ecliptic where superior conjunctions (Venus being beyond the Sun) and inferior conjunctions or loops (Venus in front of the Sun) occurred. Thus we get a pentagram. This weaving pattern is quite distinct. Certainly, the triangle of the great conjunctions of Saturn and Jupiter is an integrated geometrical symbol. But the pentagram is a weaving symbol. It is a symbol of integration in a more thorough sense. For instance, it is connected with the balance of substances in the human organism. Also, it works in the function of the glandular system. By secretion, so to speak, it establishes a balance in the body, and

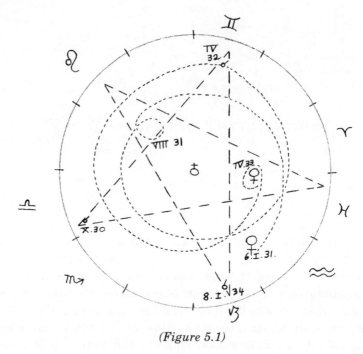

(Figure 5.1)

even works as a balancing power between the cosmos and the earthly forces. Apart from physiological functions, its rhythms are associated with the Mysteries and initiation. This too is a matter of integration in a much higher sense. We can receive

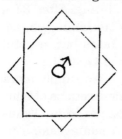

(Figure 5.2)

initiation when we are integrated as self-conscious beings in the universe, when we have found our right position in the universe. Therefore Venus is in a much deeper sense connected with the Mysteries. Above all it is, with regard to its rhythms, and so on, an image of the second half of Earth evolution. Thus it is distinct from Mars, which is an image of the first half of Earth evolution. Mars works along the pattern of the square or 90-degree angle (Fig. 5.2). Against this, Venus insists on loving integration, which is expressed in the weaving pattern of the pentagram.

First, we shall consider chiefly the three events within the Three Years and later the two events outside them. All the time we should bear in mind that they are associated with the Mysteries, the old Mysteries, the new Mysteries. The old Mysteries, we shall see, had come to an end. Through the Christ Event the new Christian Mysteries were instituted. Thus was the inferior conjunction in Leo associated with the beheading of St. John the Baptist. We read about it, for instance, in Mark, Chapter 6. What did it mean? The loop (inferior conjunction) of any planet, particularly of Venus and Mercury, signifies the termination of a cycle of individual or historic evolution. Such a cycle rises out of a loop, moves through the following superior conjunction, and comes to an end in the succeeding loop. St. John the Baptist was, in the words of Christ Himself, the greatest of all human beings (Luke 7). Once the disciples asked the Christ: "We hear the scribes say that Elijah was supposed to come again before the Messiah." He replied: "Elijah has come again; he was John the Baptist" (Matthew 11:14, 17:10-13). In other words, in John the Baptist was incarnated the individuality who was also present in Elijah. Elijah was surely one of the greatest of all human beings. Earlier we demonstrated and worked over the fact that the human head is the concentrated residue of a past incarnation—of course, not in a material but in a dynamic sense. So, we imagine that all that we hear about that powerful individuality Elijah, the outstanding figure in the line of the Prophets, according to the Old Testament, was present in the head of John the Baptist. And the head of John the Baptist carried the essence of the mighty being who must have been deeply associated with the Mysteries to know of the Advent of Christ and to be able to work toward the Mystery of Golgotha. This head was sacrificed in the beheading of St. John, and we venture to say, seen from a higher level, that John the Baptist himself sacrificed his head. Herod had John imprisoned because he had reproached Herod for his deeds. Behind Herod worked the ancient mysteries, particularly those of Phoenicia. His wife Herodias and her daughter Salome the soul-element, so to speak, in the environment of Herod, were of Phoenician descent. While in prison, John did not recant, and eventually the two women asked for his head. Herod had to give in; he had promised Salome a gift, whatever she asked. Thus John the Baptist was beheaded, and his head was carried into the hall on a platter. This gives us an idea where the ancient Mysteries had arrived. They had become utterly decadent. We are confronted here with the picture of the black counter-Grail. The Holy Grail is the dish that carries the Holy Host, the healing and ever-nourishing Spirit substance. Against this, the head on the platter is the imagination of Black Magic Art.

In the head of John the Baptist were sacrificed the ancient Mysteries, which had run their course. In Elijah, who was present in John, the ancient Mysteries had reached their final culmination, the last evening glow. John the Baptist, the Forerunner, spoke of the coming of Christ. He prepared the people who came to him for this coming by baptizing them. And though he baptized them with water, he said that the One coming after him would baptize with fire. And at the moment of the Baptism of Jesus, he saw that the promise was fulfilled. He called out: "Behold the Lamb of God, which taketh away the sin of the world" (John 1:29). He then realized that indeed the Lamb of God, an occult expression which means the Head of the Kyriotetes, of the Hierarchy of the Spirits of Wisdom, had descended into the body of Jesus. His task was fulfilled, all that which he had done as Elijah, and even earlier, in preparation for the Christ Event. The termination and decadence of the ancient Mysteries came home to him in the painful experience of his beheading.

This development culminated in the loop of Venus in Leo. The symbol that we use for Leo indicates a movement of involution from the periphery and a concentration in a little circle. Thus Leo leads us from the wide spaces of the universe down to the Earth and into ego-experience. At the time of Christ, the ancient Mysteries, the means of integrating the human being in the Mysteries of the cosmos, had come to an end. The human being now had come to live in the ego. This was the great crisis of Leo. It found expression in that loop of Venus and in the simultaneous events. In connection with the following conjunction, we shall indeed see that this was also associated with Phoenicia and with the decadence of the ancient Mysteries. This next conjunction of Venus was a superior one. It happened on May 30, 32, in Gemini. This was soon after the feeding of the five thousand. The sequence of the description of the feeding of the five thousand (the sequence must be regarded as an important key for the comprehension of the esoteric meaning of these events) in the Gospels of Matthew (Chapter 14), Mark (6), and Luke (9) indicates that the spirit of St. John the Baptist was connected with this stage of Christ's manifestation during the Three Years. The apostles were thus able, through John's spiritual presence, to make the preparation for the feeding. Christ tells them to gather the multitude, to seat them, and to provide the food, an imagination, in fact, of what must happen during the Age of Pisces if humanity is to survive. The two fishes are an indication of the two fishes of the constellation Pisces, which inspire the present age. The five loaves are an imagination of five constellations of the zodiac that are particularly connected with the human being. John the Baptist, in cosmic space after his decapitation, is able to help the apostles to provide the cosmic food that the present humanity of the "Fifth Age" and future humanity will need, in which it must find a real, spiritual nourishment: Pisces, Aquarius, Capricorn, Sagittarius, Scorpio, and Libra.

The second Venus conjunction associated with the decadent Phoenician Mysteries is found in the following chapter in Mark (7). We hear of the only occasion, according to the Gospels, of the Christ, departing from the Jewish territory and entering the land of Phoenicia: "From thence he arose, and went into the borders of Tyre and Sidon, and entered into an house, and would have no man know it: but he could not be hid. For a certain woman whose young daughter had an unclean spirit, heard of him, and came and fell at his feet: The woman was a Greek, a Syro-phoenician by nation; and she besought him that he would cast forth the devil out of her daughter. But Jesus said unto her, 'Let the children first be filled: for it is not meet to take the children's bread, and to cast it unto the dogs'. And she answered and said unto him, 'Yes Lord: yet the dogs under the table eat of the children's crumbs'. And he said unto her, 'For this saying go thy way; the devil is gone out of thy daughter'. And when she was come to her house, she found the devil gone out, and her daughter laid upon the bed." It happened in Syro-Phoenicia, just that domain where the ancient Mysteries had survived in a very decadent, one might say in a black magic, form. The woman herself was Greek and Syro-Phoenician by nationality, an indication of an accumulation of all that existed in that age as ancient heritage. Her daughter, her offspring, is possessed by a demon. It is a procreation of that dark, decadent side of what had become of the ancient Mysteries. In the conversation that takes place between the mother and the Christ, the woman is, as it were, tested. Those ancient Mysteries were often carried by a sense of arrogance, pride, and passion for power, because in becoming decadent and sliding into the wrong hands, the necessary and strict moral catharsis was more and more neglected. The sense of great humility in the woman, when she said, "yet the dogs under the table eat of the children's crumbs," had, in a sense, already cast out the devil who had possessed the offspring of the ancient Mysteries. At least, it facilitated it. For this healing the Christ had to go outside of the Jewish territory, to Phoenicia, where the ancient Mysteries were still extant, though decadent, in connection with the cults of the Gods "Baal" and Astarte, or Ashtoreth.

From the heliocentric point of view, we get still more information. The conjunction of Venus of which we are speaking occurred close to the nodal lines of Venus and Uranus. Also the perihelion of the Earth was close by. The association with the sphere of Uranus is an additional indication that this event is concerned with the Mysteries. The essence of the classical planets—from Saturn down to the Moon—is incarnate in the physical form of the human being. Uranus is already "above" the head, in the aura, invisible to sense-perception. Therefore it is the gateway to the occult, connected with initiation and with the Mysteries. The perihelion of the Earth, the point where the Earth comes closest to the Sun in its yearly orbit, is associated with the "Fall of Man" in Paradise. We have already referred in chapter 1 in detail to the perihelion-aphelion element, or line of apsides, and their manifestation in the human form. As a consequence of the Fall the human being then developed a corporeally polaric entity, head and limbs, and developed into a being that carried within the two trees, the Tree of Knowledge as head-formation and the Tree of Life, all that which is connected with the metabolism and with the limbs, particularly with the sphere of propagation. The two poles were separated. Actually we hear in the second chapter of Genesis (3:22–23): "And the Lord God said, Behold, the man is become as one of us, to know good and evil: and now, lest he put forth his hand, and take also of the tree of life, and eat, and live forever: Therefore the Lord God sent him forth from the garden of Eden, to till the ground from whence he was taken." It did finally happen, when the Mysteries became decadent, particularly in Phoenicia, that "man put forth his hand, to take from the Tree of Life, and eat," using that fruit for purely egotistical purposes.

The nodal lines of Venus and Uranus actually coincided toward the end of the fourteenth century A.D. Shortly before, when they were almost identical, in 1314, there was a conjunction of the two planets Uranus and Venus in the descending nodes, which is extremely rare. This was just a few days prior to the final extinction of the Order of the Knights Templar. On March 18, 1314, the last members of the Order, together with Jacques de Molay, their last Grand Master, were burned at the stake. The Templars had carried, on the one hand, the tradition and Mysteries of Grail Christianity. On the other hand, they also had a kind of "commercial" connection with the Asiatic Middle East. They had many castle-fortresses in Palestine, Syria, and so forth. Thereby they established "international" trade connections, and became the first bankers. They had tremendous possessions and riches, which excited in a certain moment the demonic egotism of King Philip le Bel of France. Once upon a time, kings were initiates, but later on this ceased to be so. Philip le Bel was certainly not an initiate. He was more voracious than the average human being. And he decided to destroy the Order of the Knights Templar in order to acquire their treasures. So deep had the last faint shadows of the ancient Mysteries sunk. The demons who had long before invaded their holy precincts fought the new Christian Mysteries, successors of the Holy Grail, with all available might. They worked with the forces in the human being that were associated with the aphelion of the Earth. These forces were liable to combine in the human being with passion, lust, and the tenacity to hold on with all possible means to the perpetuation of the physical-material element.

However, this need not be accepted as the inevitable fate of the ancient Mysteries. In order to get some clarity into this we will use some more cosmic information. The perihelion of the Earth and the node of Uranus coincided in 382 B.C. It was the time of Socrates and of Plato. Socrates died in 399 B.C., just a few years prior, and Plato in 347 B.C. In order to get an idea of what happened then we had best study Rudolf Steiner's *Christianity as Mystical Fact*. In Chapter 4 we find an illuminating description of Greek civilization at that time. The Greeks transformed harmoniously the experience of the ancient Mysteries into philosophical teaching. So, the Greeks effected the transition in a healthy fashion.

We come to the last loop within the Three Years. The inferior conjunction took place exactly at the vernal point, just a few days prior to the Mystery of Golgotha. At the time of Golgotha, Venus was actually still retrograde. So it was still in the loop. This was connected with the raising of Lazarus. The raising of Lazarus became one of the foundation stones of the new Christian Mysteries. Lazarus was John the Divine. He probably had some association, even if it was not the same individuality, with the writer of the Revelation of St. John, or the Apocalypse. The latter is indeed a manifestation of esoteric Christianity. We hear that Lazarus was supposed to have died and lay in the grave for three days. This is exactly what happened in an ancient initiation. In Egypt, for instance, the neophyte lay in the sarcophagus for three days and was then raised. Thus, we must see the raising of Lazarus as an initiation. There is, however, one difference. On this occasion the initiation was performed openly. All the world, whoever was present, could witness it. Initiations in the ancient mystery places took place in strict seclusion. All the so-called profane human beings then had to leave the temples. And now here in the case of Lazarus an initiation was performed in front of all the world. This was the reason why the Pharisees came to the decision that this man Jesus had to die because he had betrayed the Mysteries, and violated the severe oath of secrecy pertaining to the Mysteries. A traitor had to die. This was the law and this was the reason for the decision to have Christ Jesus crucified. This event and the openness with which the Christ performed the initiation describe the character of the new Mysteries. They are "open" and yet secret. They are open in the sense that there is no external closing of the doors any longer. The protection of the esoteric element lies in the spiritual understanding of the human being. Only such a spiritual understanding can penetrate to the inner realm. This is the change that was introduced as the road to the new Mysteries.

We come to the two conjunctions of Venus outside the Three Years. The first one was in October, 30 A.D. To understand this it is advisable to study Rudolf Steiner's lectures on the "Fifth Gospel." They give a description of the events leading up to the Baptism. Jesus was associated with the Essenes, the order of the Essenes, who had settlements, for instance, near the Dead Sea. They lived a secluded life of extreme purity and abstinence. Thereby they tried to keep out the evil influences. They even insisted on using their own special gates into the cities. These gates were impenetrable to the adversaries, Lucifer and Ahriman. Jesus had a terrible experience at one such gate. He saw that Lucifer and Ahriman were indeed rejected from them. But he also realized that they then fell upon the rest of humanity. He saw that the Essenes bought, so to speak, their purity and their holiness by putting the burden of evil on the humanity outside their order. And so Jesus realized that this was no longer the way to salvation for humanity. He was shattered by this discovery and went back to his mother. They had a deep conversation. After that, with his last strength, he went down to the Jordan and was baptized by John the Baptist. This was for Jesus a kind of death experience because he offered up his body to the Christ. He as an Ego withdrew. All this happened on or before the 6th of January, 31 A.D. So, we presume that these events happened at the time of that superior conjunction, or in its aftermath. It occurred in the constellation Libra, almost at the ingress to Scorpio, which is another characterization. Indeed, we see Lucifer and Ahriman, who were fleeing from the gates of the Essenes and falling upon the rest of humanity, coming back after the Baptism and the forty days in the desert. They came to tempt the Christ. They knew that Jesus had recognized their presence earlier. Now they made a last attempt to frustrate once and for all this deed of the spiritual world to save humanity from their attacks. So all this one would expect to be present in this Libra-corner of the pentagram.

After the Three Years, about three-quarters of a year after Easter 33, we come to the event at the last, fifth, corner of the pentagram. It was a superior conjunction of Venus and happened on January 8th, in the year 34. It was in Capricorn, which was in Greek mythology considered to be the gate to the divine-spiritual world, opposite Cancer, building the bridge and opening the doors. Hercules ascended through it to the heavens after his death. What happened then? In the Christian Calendar we find the 25th of January dedicated to St. Paul at Damascus, that is, the Conversion of St. Paul at the city gate of Damascus. Paul, or Saul as he was then still called, was present at the martyrdom of St. Stephen who was the first Christian martyr (Acts 7, 8). The latter's death is commemorated in the Christian calendar on the 26th of December. After that Saul went, so to speak, as the authorized persecutor of the Christians to Damascus. He did this because he was convinced that the cause of Christianity was detrimental, particularly to the Jewish tradition. First, there was the accusation against Christ Jesus that He had betrayed the Mysteries. Then there was the miserable death on the cross, which could not evoke the confidence of Saul that this was really the expected Messiah, who would free the Jews from the Roman yoke, and so forth. This was impossible— it could not be. And then there was the claim of the Christians that He, the Christ, had risen from the grave. All this was too much for Saul, who had been brought up, so to speak, in strict Hebrew esoteric tradition. Therefore he felt urged to persecute the Christians. After the execution of St. Stephen he had, according to the Acts of the Apostles, papers made out by the councils in Jerusalem so that he could go to Damascus, to continue there with the work of eliminating the Christians. And then at the gate of Damascus he had that shattering experience ("Saul, Saul, why persecutest thou me?"), his conversion. All this happened very probably in the neighborhood of the last, fifth Venus event along the pentagram. Now, the new Christian Mysteries were reestablished. This last event laid the foundation for a new experience of spiritual reality by the ever-possible awareness of the Presence of the Risen Christ, who had confronted St. Paul and convinced him of the truth of Christianity. The conversion of St. Paul prepared the Second Coming, the manifestation of the Christ in the ether body. Thus we can also understand that these things are essential for our present age. The pentagram of Venus is of a perpetual nature. For instance, we find that the conjunction of 34 A.D. was followed after eight years, in 42, by another similar event in the very same corner, that is in Capricorn. Likewise the loop of 33 was followed by another loop in 41, close to the vernal point. Thus all the five points were repeated. However, we shall always notice a slight difference. They all fall back between two and three degrees in the course of eight-year intervals. It doesn't seem much, but it makes all the difference. Over centuries and longer times it simply means that the pentagram is very slowly turning backward in the ecliptic. Thus we can say that, for instance, the inferior conjunction of 33 A.D., around the time of Golgotha, has at present arrived, after a complete rotation in between, in the constellation of Leo. About the beginning of the century, in 1903, it had entered Leo, coming from Virgo. As an inferior conjunction it recurred in that constellation in 1911, 1919, 1927, 1935, 1943, 1951, 1959, and 1967. During the first half of the next century, it will enter the constellation of Cancer. Likewise, the other four points of the pentagram move backward through the zodiac.

A study of these events and these rhythms can bring home to us that our contemplation of the happenings in the heavens during the Three Years are not only of historic interest. By their rhythmic return an activating commemoration is effected. And in this we see the significance of our studies. We can thereby gauge where we stand in history, what kinds of facilities are offered thereby. By no means do we suggest that things must happen in the sense of unalterable fate. Actual experience and work with these rhythms refute this idea. But a gentle cosmos offers them as tools, so to speak, for use by the spiritually free human being. Of course, their refutation is not without consequences, not in the sense of any kind of revenge, but rather as denial, and eventually, a dissolution of self.

6.

The Movement of
Mercury during the Three Years

IN THE PRESENT CHAPTER we shall work out details concerning Mercury. This will be somewhat more complicated than the other planets. It is particularly the rhythms of Mercury that interest us. In the course of one year it performs three complete cycles. Figure 6. gives the heliocentric point of view. The Sun is in the center. Mercury moves around the Sun. Beyond it is

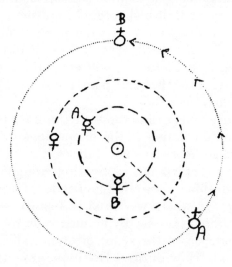

(Figure 6.1)

Venus. In the outer circle we see the orbit of the Earth. It can happen that at one time the Earth is in position A and Mercury in point A of its orbit. This is a superior conjunction (or heliocentrically an opposition), Mercury being behind the Sun, far out in space, as seen from the Earth. Both move on, and in the course of 116 days the Earth moves into about the position B. Mercury completes an orbit once in 88 days and comes back to position A. During the remaining 28 days (up to 116 days), it goes still further and moves into B. We find it again in superior conjunction with the Earth. This we call a cycle. From the geocentric view this presents itself somewhat differently. For this we need to look at Figure 6.2. There the Sun is moving around the Earth and we visualize the Sun in position A and Mercury behind it, as seen from our planet. From there Mercury goes

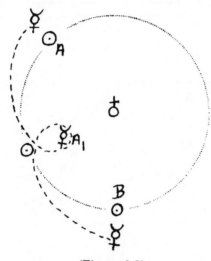

(Figure 6.2)

around the Sun, moves on, and finally finds itself in position B, again behind the Sun. Thus it has completed a cycle. But something happened midway in the cycle. Mercury, on its orbit around the Sun, would have arrived in this position A1. It is obviously in front of the Sun, seen from the Earth. This movement appears from our standpoint, according to the laws of perspective, as a loop. Of course, Copernicanism will say that this is an optical illusion, caused simply by the relative positions of Sun and Earth and planet. Nevertheless, we take it as a significant cosmic symbol, which vividly describes what is happening in such a moment. In more than a metaphorical sense we can

say that in position A Mercury gathers up cosmic substance
from space beyond the Sun. After that Mercury goes into a
phase as evening star, though it may be difficult to observe on
account of its smallness and nearness to the Sun. (Copernicus
never in his life saw Mercury.) After that the planet moves in
front of the Sun and comes much closer to the Earth. One can
say that it hands over to our planet what it had gathered in
cosmic space, while it was behind the Sun. This is, however, not
the end (Fig. 6.3). Again Mercury moves on, into B, the next
superior conjunction, followed by a loop in B1, a superior conjunction in C, and finally a loop in C1. After that Mercury
moves back into, approximately, the position A. These are the
three cycles of the planet in the course of one year.

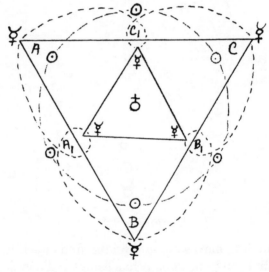

(Figure 6.3)

It is obvious that we thereby get two triangles, whose corners
are indicated by the superior and inferior conjunctions. However, on account of the relative distance of the planet from the
Earth, the inner triangle of the loops is much smaller than the
triangle made by the superior conjunctions. They are not quite
equilateral, which is due to the apparently varying speed of
Mercury. When it is moving through its perihelion it appears to
move fastest, whereas in its aphelion it appears slower on

account of its far distance from Sun and Earth. The movement from one superior conjunction to the next takes about 116 days, we said. Three times 116 days makes 348 days. So there are still 17 days left in one year of 365 days. Thus it happens that the triangles constantly shift backward in the ecliptic. For instance, it takes about six years for the point A (Fig. 6.3) to get into the position of C, and of course, the others fall back correspondingly. It takes about twenty years—that is, not quite three times seven—for the corners of the triangles to come back to the original positions. Thus this double triangle, or hexagram with some imagination, moves rather fast compared with that of Venus. The movement of the Venus pentagrams takes about 1,200 years to go through the whole ecliptic. The geometric patterns that the planets inscribe in the heavens are, with reservations, the real stars. What we see in the heavens is, in a sense, only the pen that indicates the real star.

If we take the hexagram of Mercury realistically, what does it convey to us? It comes close to the symbol of Figures 6.4 and 6.5. It means that the big triangle is always connected with certain cosmic substances or elements that are being collected. Mercury is associated with intelligence. In the sense of the bigger triangle, this would be cosmic divine intelligence, because there the planet is right out in cosmic space. The small triangle would then be an image of cosmic intelligence being received on Earth, for instance, into human intelligence. We could translate this into the attempt of an imagination in Figure 6.5, the greater triangle depicting a cosmic being who spreads his cloak over the smaller being, represented by the inner triangle. The latter lifts up his arms, as it were, in a gesture of supplication and reception. This can give us quite a good idea of the nature of Mercury, of how it is connected with the background of human intelligence, and how this intelligence

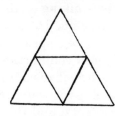

(Figure 6.4)

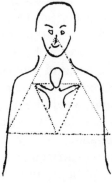

(Figure 6.5)

works into the will and can perform deeds out of a realization of cosmic intelligence. Thus Mercury is always associated with this twofold action, which is also expressed, for instance, in human breathing. We take the air that is all around us in space into our organism by inhalation. We transform it in the body, and then we exhale it again, of course, in a different composition. In this sense, the Mercury cycles play a most important part in the events of the Three Years.

During the two and one-third years, between the Baptism and Golgotha, we find seven cycles of Mercury. Each one displayed the gesture of going far out into cosmic space, beyond the Sun. The inferior conjunctions, the loops, signified a bringing down of cosmic forces into Earth reality, into earthly deed. Thus, for instance, it might not appear to be farfetched if we connect these seven rhythms with the so-called "Signs" in the Gospel of St. John. In the Gospel of St. John we hear of seven great deeds of Christ. The first is the wedding at Cana, right at the beginning, then the healing of the nobleman's son of Capernaum, the healing of the sick man at the pool of Bethesda, the feeding of the five thousand, the vision of Christ walking on the sea, and the sixth, the healing of the man born blind. The seventh was the raising of Lazarus. As far as one can speak of chronological evidence, it appears that the superior conjunction was connected with a kind of gesture: the gesture of reaching out into the cosmos, of using Mercury as a tool, as it were, to gather up certain cosmic substances, or ingredients, and then to bring them down into one of the seven signs at the time of the inferior conjunctions, or loops.

This is described from another angle in the Gospel of St. Mark. In the first chapter (32 and 39) we hear that in the evening many sick people came to the Christ and were healed. And in the morning He went out to a solitary place and prayed. We can well imagine what it meant when Christ prayed. He brought down cosmic forces, as an ordinary man would pick up his tools. He is the great Ego, the Self of the cosmos, Who can use the ingredients of the cosmos in independence. By no means do we suggest that it was a matter of the Christ's being dominated by the stars. Rather it is a matter of being in command of the cosmos, as the ego of the human being can be in command of the body. Thus He reached out into the universe, and then brought the ingredients down onto the Earth, to use them for His deeds. According to the Gospel of St. Mark, the healings in the evening were made possible by the morning forces brought down to Earth in praying. This is a reality. The descriptions in the Gospels are by no means belletristic para-phrases. They carry tremendous reality, which must from now on into the future be taken very seriously if humanity decides to follow Christ.

We venture to mention in this connection the experiments and achievements of Dr. Hauschka in Germany. He discovered that indeed the morning and evening forces can have a tremen-dous influence on substances. Normally, substances like plant juices and also medicines in liquid form decay because of the natural chemical reaction in liquids. Therefore alcohol or chem-icals have to be used in order to preserve substances, which is not always desirable. So, Dr. Hauschka looked for different means of preservation. He came to realize that when he exposed such substances, juices or medications, to the moments of the rising and setting of the Sun, shielding them during the rest of the day against environmental impacts, that he could preserve the liquids up to seven years. This was a remarkable achievement. He used the power of the Sun. In con-nection with another context we came to the conclusion that the Sun, although it ultimately dissolves space and the ele-ments therein, nevertheless creates space by the suction it exerts on its environment, deep within the cosmic-sidereal spheres. Thus, as much as it seems to dissolve substances, for instance on its own surface, indirectly it has an effect on the substances on the Earth. This is accelerated at the moment of the rising or setting of the Sun. With regard to the Moon, a dif-ferent effect is probably connected with rising and setting. The Moon is, among all its other correlations, also associated with growth, wherever life becomes manifest in the kingdoms of nature. Thus all the planets establish their particular func-tions, possibly in connection with such rhythms as their rising and setting. In this way, the forces that the Christ brought down to the Earth, were cosmic forces, for the sake of healing and redeeming the consequences of the Fall. Sooner or later we will have to understand and to apply such principles for the sake of our survival.

At present we still rely on nature and the cosmos to provide us with everything that we need in order to conduct our life on this planet. But we are slowly coming to realize, partly by gigantic catastrophes in the making, for instance, concerning nutrition and many other things, that we must learn to find new standards of existence on our planet. It becomes more and more obvious. The pollution in all spheres of the Earth—soil, water, and particularly air—will reach such a pitch that we will not be able to rely, as a matter of course, on that which nature has given us, so far anyway. We will have to learn to do the jobs ourselves and not solely rely on and passively accept what nature and the cosmos have given to us so graciously up until now. Otherwise, one day, we may no longer be able to incarnate as we have up to the present. When we incarnate, we go through an embryonic process. We receive from nature the material and from the cosmos formative forces. Out of the combination of the two, the body in which we dwell for a lifetime is built up. This may well come to an end. For instance, Rudolf Steiner has indicated such developments. He spoke of times to come when the present kind of propagation of the human race will cease. By then humans will have learned new ways of associating spiritually with the world of matter in order to build up and conduct life on this planet. That which humans of the future must evolve in themselves as spiritual faculties was set out in great archetypes, or examples, by the events of the Three Years. So, looking at Mercury we must always bear in mind that the taking up of cosmic intelligence, divine intelligence, and bringing it down into earthly deeds of healing and redemption by the Christ was not done for the sake of producing miracles. These deeds call for the eventual "Imitation of Christ," of which Thomas à Kempis, the medieval mystic, spoke.

In order to clarify matters we will draw two diagrams containing the seven cycles of Mercury during the Three Years (Figs. 6.6 and 6.7). About forty days after the Baptism (which occurred on January 6, 31) a superior conjunction took place at the ingress into Pisces (Figure 6.6). After that, Mercury moved into a loop in the constellation Aries. That was around Passover, 31 A.D.

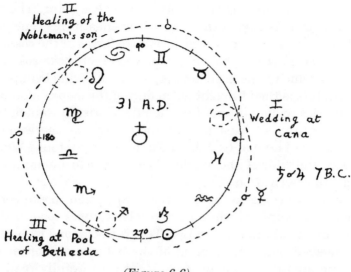

(Figure 6.6)

Thus Mercury started just at the transition from Aquarius to Pisces with a superior conjunction. Now we will imagine that through this the Christ reached out with a mighty gesture to receive from the cosmos divine intelligence from the direction of Aquarius and Pisces, and that He brought it down at the time of the loop in Aries about Easter of that year. This would stand behind and be the cosmic source of, the first Sign, the wedding at Cana. Water was then transformed into wine. This "miracle" was enacted out of the secrets of heavenly and earthly "water." The wine is an indication that this was an ego-provoking deed, which it still was in those days. By knocking, so to speak, against the organism of the body, it could still bring down the ego into a human being, which it no longer does. The power to do this came from the Aquarius-Pisces region, above which we find Pegasus, the constellation of Pegasus, the Winged Intelligence, as against earthbound, stagnant intelligence. Above Pisces is the constellation of Andromeda, an image of the soul of humanity that has been elevated to cosmic heights. In this we can discern the background of the conversation between the Christ and Mother Mary at the wedding. (John, 2:3-5. The translation should, according to Rudolf Steiner's information, be: "Woman, what does there weave between thee and me?")

Heavenly Virgin-forces of "doing" enter His being. The "water" of Aquarius is part of the Great Sea beneath this constellation. It is an image of the etheric formative forces of the cosmos. The water on Earth is the carrier, or vessel, of these cosmic ether forces, which regenerate and replenish all living beings.

All this is then brought down during the loop of Mercury into the Aries region. In that first sign the Christ presented Himself to humanity, at least to that small circle at the wedding, as Aries, the Lamb of God, the head of the Kyriotetes, Who takes away the "Sin of the World," Who offers to humanity a new spiritual activation.

Next we come to the superior conjunction in the constellation of Gemini, which is then resolved into a loop in the constellations of Cancer and Leo. This is the second loop, which is connected with the healing of the nobleman's son at Capernaum according to John 4:46-54. If we carefully read all the details of that healing—every word is significant—it can give us an idea of what happened, of how the Christ brought down forces from the region of Gemini. This region was then still an expression of the cosmic archetypes of the prevailing hierarchical order of the human community. In connection with the fact that it is a nobleman's son, the hierarchical social order is especially emphasized. This event in John, together with the redeeming humility that rises up in a centurion, is connected with a similar story of a healing, also at Capernaum, described in Matthew, 8:8-13. The corresponding loop took place in Leo, which symbolizes a development from the periphery to the center-point, to the ego that must be evolved on Earth and whose advent must become the sole criterion of social order.

We come to the next superior conjunction in Virgo. Virgo is a rather elongated constellation. It reached at the time of Christ across the point of the autumnal equinox into the area of the ecliptic sign of Libra. This superior conjunction was then resolved in a loop at the point of transition into Sagittarius. This can be associated with the healing at the pool of Bethesda (John 5:1-16). A festival took place, and the Christ went up to Jerusalem, and there at the pool he found the man who had had an infirmity, as it is said, for thirty-eight years, and he was healed. Why is the fact that he had had an infirmity for thirty-eight years especially mentioned in the Gospel? There must be a meaning in this, because in the Gospel's every detail, every word is significant. These thirty-eight years contain deep mysteries of the Christ Event. The healing happened in 31 A.D. If we go back thirty-eight years from there we come to 7 B.C. In 7 B.C. the "Spiritual Nativity" of the Jesus of whom the Gospel of St. Luke speaks took place. There is the possibility, as far as we can see, that it was also the "Spiritual Nativity" of the Jesus of the Gospel of St. Matthew. The Jesus according to St. Luke was born just a few days prior to the turning point of time from the B.C. to the A.D. era. This means that he was born at Christmas of the year Zero in an astronomical sense, or 1 B.C., in the calendar sense. For astronomical calculations we must bear this in mind. If we are concerned with a date in 7 B.C., we take the zero position minus six and a fraction of a year. Likewise, a date in 31 A.D. is thirty plus a fraction of a year away from zero. The sum total is then thirty-seven years. This does not seem to agree with the thirty-eight years of the Gospel. However, with regard to these indications in the Gospels, we must take things from a different angle. For instance, we hear of the three days that passed between Golgotha and Easter Sunday morning. He rose after three days it says. This does not mean that it was three days in quantity of time. It was a little more than thirty-six hours from Good Friday afternoon until sunrise on the following Sunday. But we must see it as events on three consecutive days, or events during thirty-eight consecutive years, and so on. In this sense the "Spiritual Nativity" took place six years and a bit more before zero. Thus we come into the year 7 B.C. according to the calendar.

Now, we have slipped in a totally novel concept, the "Spiritual Nativity." This seems very complicated and confusing. However, the world and its inhabitants are not as simple as some people may want to have them. We are used to looking at nativity as the moment when a human being enters the physical world by birth. This is certainly a significant moment, but that does not exclude the other events around the incarnation, which may be equally if not more important, particularly if one realizes the fact of reincarnation and the existence of the soul before birth. In this sense the "Spiritual Nativity" is, so to speak, the spiritual birth of the human being in the cosmic world, while possibly still years away from incarnation. This "Spiritual Nativity" can therefore take place years before birth, or years after birth. As it is a "spiritual" birth, connected with the higher being of the human being, which only shines, as it were, from cosmic heights into the physical incarnation, it is not bound to the laws of physical space and time. However, there exist ways and means of finding this event as a mathematical-astronomical fact, connected with the physical nativity. For this purpose we take the Moon at the moment of birth, for instance, of the birth of Jesus according to the Gospel of St. Luke. We regard that Moon as the last stage on the descent into incarnation. The Moon is closest to the Earth; therefore it indicates the last step down onto the Earth. Next we take the Moon node and calculate the time when the latter was standing in the place of the Moon at birth. Thus we find the "Spiritual Nativity," which Rudolf Steiner describes in *Human and Cosmic Thought,* a lecture cycle given in January 1914. The calculation of the "Spiritual Nativity" of Jesus leads us back to one possibility, that is, in the year 7 B.C., and we have good reason to accept this one date as being appropriate.

This point in history gives us very interesting and pertinent perspectives. This was apparently one of the signals for the Magi, the Three Kings who came to visit the child (Matthew 2). We have come to the conclusion that the star of the Magi was not one of those tiny twinkling things in the heavens. The Magi were initiates and astrologers in the noblest sense. The traditions of the Mystery schools had taught them to watch out for such events as a great conjunction like the one that occurred in 6 B.C., and more. That was the first signal, and by their wisdom of the smaller cosmic cycles they could know when the birth of Jesus would take place. The great conjunction of 6 B.C. occurred in Pisces, which, in one sense, is as much as saying, "The time is fulfilled." Pisces is the grand finale of the twelve constellations; Aries goes in front as the leader of the heavenly host. In the image of Pisces, which is connected with the cosmic archetype of the feet, the human being comes down to stand in the reality of Earth existence.

The great conjunction had also played into the life of that man who had been sick for thirty-eight years. Through some circumstances, he must have known of the great prophecy connected with that star event, and he hadlived since that time in expectation of the great redemption, of the Messiah to come. He could not find him, and that was the deeper cause of his sickness. In the moment in which Christ stood in front of him, a remarkable cosmic rhythm came to completion. We mentioned earlier that the event of 6 B.C. in its function as the "Spiritual Nativity" of Jesus was associated with the position of the Moon node at that time and its correlation to the Moon at Jesus' birth. In 31 A.D., after 37.2 years, the Moon nodes had returned, after two orbits, to the original position. The Moon nodes are openers of the gates to the higher spheres of the cosmos, gateways from the Moon to the Sun sphere, or the astral world. Thus that moment in 31 A.D. for the sick man was an important stage of his long life of expectation. When he was healed he did not yet realize that the Messiah was standing in front of him. Only later, when Jesus found him again in the temple and said to him, "Behold, thou art made whole: sin no more, lest a worse thing come unto thee" (John, 5:14), does he recognize Him. The sin was the incapacity of spiritual recognition of the Greater Ego of the cosmos. This is a typical Sagittarius situation. We remember that the healing took place in connection with the loop of Mercury in that constellation. The aiming of the centaur with bow and arrow, and the tremendous evolution indicated by the horse body being resolved into the human form, are vivid imaginations of this.

We come now to the fourth cycle of Mercury. That would then be the year 32 A.D. (Fig. 6.7). This one was somewhat similar to the first one in 31 A.D., only it had fallen back by a few degrees. The corresponding superior conjunction was in Aquarius, followed by a loop in Pisces, close to the ingress into Aries. It was shortly before the Passover of 32 A.D. We read about the events that took place then in John 6:1-15, and also in the other Gospels. It was the Feeding of the Five Thousand. Again, the region of cosmic sustenance, Aquarius, was involved, similar to the Changing of Water into Wine, and also Pegasus and the constellation of Cygnus, the Swan. The presence of the

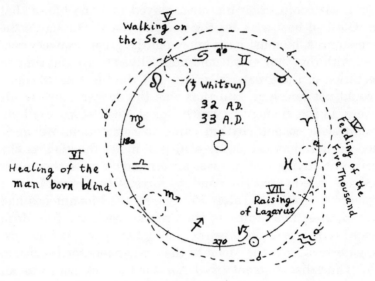

(Figure 6.7)

Spiritual World is indicated in these constellations. Out of those regions the Christ brings down cosmic forces at the time of the loop that takes place in Pisces, where the vernal point was to be in following centuries. One year earlier it was the wine. Now five loaves and two fishes are multiplied to feed five thousand. It was a feeding with nourishment. The two fishes are those of the constellation Pisces. To this is added Aquarius, Capricorn, Sagittarius, Scorpio, and Libra. It all referred to the future of humanity. At the time of Christ the vernal point had just about passed over, astronomically speaking, from Aries into Pisces. The future will see the vernal equinox in Aquarius, Capricorn, and so forth. Thus the feeding implied feeding a future humanity, first that of the Age of Pisces, the Fifth Post-Atlantean Age. It is a great Imagination, which contains not only pictorial conceptions but concrete spiritual realities, which a future humanity will have to live up to.

We move on to another superior conjunction in Taurus, followed by a loop in Leo. This is associated with the fifth Sign, the vision of the apostles of Christ Walking on the Sea (John 6:15-21). According to the description in the Gospel, this appears to have taken place immediately after the feeding of the five thousand. However, in the sense of sequence in the Gospel, this need not be taken as that. There is a strange coincidence reported in three Gospels, St. John, St. Matthew, and St. Mark: When they saw Him walking on the water, they were afraid, saying "It is a spirit." Why should they be afraid? Following the "Feeding of Future Humanity," they saw Him as He would appear to that future humanity, in an etheric, in a spirit, form. They experienced Him as He actually appeared to them after the Resurrection. This was a manifestation of Leo, and Saturn entered Leo later, after Golgotha. The apostles had a pre-vision of what was to come. The preceding superior conjunction in Taurus indicates that Christ had brought absolute mastership of the physical, even of the material, world from the cosmos. Taurus is connected to, is an image of, the Logos, the Creative Word, who "was in the beginning with God. All things were made by him; and without him was not anything made that was made" (John 1:1-3). We note also the reference to "the walking on water," which we take as an inference to the appearance of Christ in the ether.

We come now to the sixth cycle, still in the year 32, with a superior conjunction in Virgo, followed by a loop in the constellation of Scorpio. This is connected with the sixth Sign, the healing of the man born blind, according to John 9. Scorpio is a cosmic image of the blindness from which all the human race increasingly suffered, particularly in our present age. It is also a vivid expression of the heavy arguments that went on after the man was healed, between him, his parents and the Jews (John 9:8-38). In the conversation between Christ and this man, after his arguments with the Pharisees, we can see a redemption of Scorpio (John 9:35-38). All this was done out of the cosmic essence of the preceding superior conjunction in Virgo, carried into Earth reality by the power of the Ego of the cosmos, or the Sun. This means to say that the answers to these Scorpio-queries can be found in the attainment of divine wisdom, the wisdom of Isis-Sophia, or Virgo. There can also be expected answers to such problems as heredity, and so on, which are implied in the story (John 9:2-3).

Finally we come to the year 33 A.D. and to the seventh cycle of Mercury. There was first a superior conjunction in the constellation of Capricorn, which descended into a loop in Pisces. This is connected with the last Sign, with the raising of Lazarus, which must be regarded as an initiation. (See Lecture 5.) Capricorn is, according to ancient wisdom, the portal to the spiritual world, through which the soul of the human being proceeds at the moment of death. Christ brings these cosmic imaginations down into Earth reality. And when the loop takes place in Pisces, the "last" constellation is made by Christ the cosmic cipher of the "first" and highest, of the initiation perspective contained in the Gospel of St. John, and eventually in the Revelation of John the Divine.

7.

The Manifestation in World History of Cosmic Events during the Three Years

WE HAVE REPEATEDLY pointed out in this course that since the middle of this century, all the events with which we have been working in connection with the Three Years are with us again. They are effective in history. That this is so is due to the "Presence," to the "Parousia," of the Christ from our present age on into the future. Therefore we shall make it our task in this connection to demonstrate how the latter event is evident in cosmic chronology.

When we human beings pass over the threshold of death into the spiritual world, we first lay the physical body aside. Then, during the first three days after death we still live in our etheric body. This etheric, or formative, body is a time organism. This means that it is like a chronicle in which all the events in the life of a human being are inscribed. They are present, so to speak, side by side, although in time they may have happened apart. Thus the past life appears as a great tableau. And our souls are confronted with this tableau during the first three days after death. However, after three days this ether or tableau body is dissolved into the planetary world, because that is the world from which it was taken at incarnation. Now, we can well imagine that in the moment of death on Golgotha this did not happen in the same manner. The ether body of Christ Jesus did not dissolve. As a matter of fact, even the ether bodies of the great initiates do not, as a rule, dissolve. They may remain intact. That has happened in history over and over again, and these ether bodies may then work all the more in human history. Nevertheless, that ether body which was freed from the physical body of Jesus on Golgotha went through a tremendous evolution. It was taken up into the cosmos, that is, to the limit of the etheric cosmos. In other words, the ether chronicle of the Deed of Christ was communicated to the whole universe. But at a certain moment it returned, just as an echo would return, and since the middle of the thirties of the present century it has come back. Now it is again in the neighborhood of the Earth, in the aura of the Earth. In order to get an idea of how this worked, we must realize that we are here confronted with certain mysteries with regard to the dynamics of time. We experience time here on Earth as a definite ratio of sequence, of things happening one after the other. We make the day of 24 hours and the year of 365 days the basis of our time concepts. The day is made by the rotation of the Earth around its axis, which causes the phenomena of day and night and the interval of 24 hours. The year of 365 days is caused by the movement of the Sun, or apparent movement in the sense of modern astronomy, around the Earth and its return to the same position in the ecliptic. This takes a time of 365.25 days. Now, we can well imagine that time in the cosmos is something totally different from earthly time. There, different time ratios must be the foundation of chronology because the rotation of the Earth, for instance, does not directly concern the cosmos. The question now is, what could be a universal time ratio, and can we have access to it? Rudolf Steiner made some suggestions. In a lecture on December 3, 1916, he pointed out that in the spiritual world time is thirty times longer than here on the Earth. So, what we experience as a "year" here on the Earth would appear in the cosmos like an interval of "thirty years." Why should this be so? Time, as we said, is founded on certain rhythms of celestial entities, like the Sun or the Earth, with regard to time on the Earth. However, in the cosmos a number of different rhythms prevail. One should expect that one of the most archetypal rhythms would be one that was associated with the edge of the whole solar system. In a classical sense (Uranus, Neptune, and Pluto are "newcomers" from a certain angle), Saturn, the orbit or the sphere of Father Chronos, as he was called in Greece, may provide the ratio of time we need. This planet was conceived as Omnipotent Father Time, also as the Omnipotent Father of History. He indicated time in a divine, a spiritual sense, by his rhythms. The revolution of Saturn through the zodiac takes about thirty years, or precisely 29.4577 years. Saturn needs this much time in order to complete a sidereal orbit around the Sun, which means that it comes back to the same fixed star in the zodiac in the course of such an interval of time.

The ether body of Christ Jesus was taken up into the cosmos, and we must now imagine that it was into the time that prevails in spiritual cosmic space. On Earth it contained a time quantum of thirty-three years. If we calculate precisely, that is, if we start from the birth of Jesus according to the Gospel of St. Luke and go right up to the time of the Mystery of Golgotha, we actually have a period of 32.28 years. Golgotha happened in the course of 33 A.D. This means that thirty-two years had passed since the New Era started. To this we add the fraction of a year up to Golgotha, and also the few days immediately before the turning to the New Era, from the 25th of December to the end of the year Zero. Thus we have a time complex, a time "being" one might call it, of 32.28 years. And these we transpose into cosmic spiritual years. That means we multiply these 32.28 years by 29.4577, the figure for a Saturn "year." Thus we come to 950.895 years (Fig. 7.1). Seen from Saturn, this is 3,228 "years." Seen from the Earth, 950.895 years after Golgotha, the ether body of Christ, the time-organism of the Deed of Salvation, has arrived at the periphery of the solar system. Now it is completely spiritualized, elevated to cosmic magnificence and power. But it is far away from the Earth.

1 ♄ orbit = 29.4577 yrs. x 32.28   (Life of Jesus Christ)
        = 950.895 yrs.
    plus   33.25   = Golgotha, April 3, 33 A.D.
        ─────────
        984.145 A.D
    plus   950.895   = return-cycle (of 29.4577 x 32.28 yrs.)
        ─────────
       1935.040 A.D. (January 6th, 1935)

(Figure 7.1)

By now we have arrived at the end of the first millennium A.D. We started out from the year 33.25. The time up to the beginning of April, when Golgotha took place, would add just one quarter of a year. The 33.25 plus 950.895 years (see above) would bring us to 984.145 A.D. And now this Time Being flows back toward the Earth, as an echo comes back. This takes the same time ratio as the going out. The ether organism of 32.28 years in Saturn terms, that is, after another 950.895 years, arrives in the neighborhood of the Earth. Thus, 984.145 A.D. plus 950.895 brings us to the year 1935, or precisely, 1935.040. This is the time between the 6th of January, the day of Epiphany, and the 26th of January, the day of St. Paul at Damascus, in the year 1935. At that moment this Ether Being was back in the environment of the Earth. It has always been present in the cosmos since Golgotha, and therefore certain individualities who had gone through extreme experiences and preparations realized the Presence, that is, the Christ in his ether body. For instance, St. Francis of Assisi was in such a position. He could thus experience the Christ and even receive the stigmata. Others had similar experiences. But it was only in 1935 that it was back and close to all humanity, at least potentially. The year 1935 is precisely the date that Rudolf Steiner indicated as the commencement of the age when increasing numbers of human beings would be able to perceive the Christ in his ether garment. This will be accompanied by the development of a natural clairvoyance. The ether body of the human being, which in past millennia had become more and more submerged in the physical body, will gradually lift itself out of the head region. Thereby the human being will be able to perceive "similar by similar," to perceive Christ in the ether body through their own emancipated ether bodies. This is the reason for many of the things that manifest at present, right down to so-called psychiatric problems. More and more, human beings have been speaking of experiences like St. Paul had at Damascus. These experiences are with us and they have been noticed. But the danger is that a materialistically oriented humanity characterizes them as sickness. This may be the reason why we hear very little about such experiences. People who have them may remain silent with good reason. It needs a spiritual science to understand this development, and it also shows our responsibility. Knowing the facts, we could help people.

The ether body of Christ, which we said is close to the Earth, contains all His words and deeds during the Three Years. They are present in it as activating, living memory, not only mere reflection. Thus they will manifest—through the Presence of Christ in them—in present and future humankind. It has happened, and it can even be "proved," in a certain sense.

Whenever planets return in the course of their orbits to the positions at the time of the events in Palestine, the original perspectives of Christ's deeds become manifest. He "full"-filled the star-configurations that accompanied the Incarnation with new spiritual meaning, for the future. This is reactivated and can enter human lives, from now on into the future, as a challenge whenever planets return to the original positions, or when "transits" take place.

For instance, at the beginning of the present century Saturn was opposite the places in the zodiac that it had occupied during the Three Years. From 31 to 33 A.D. it moved through the ascending node of Jupiter, its own ascending node, the perihelion of Venus, and the ascending node of Neptune. In 1901–03 Saturn moved through the opposite positions, that is, the descending node of Jupiter in 1901, and so forth. Rudolf Steiner gave then, starting in October 1901, a significant course of lectures, which were later published in book form as *Christianity as Mystical Fact and the Mysteries of Antiquity* (full title). In the Introduction Steiner makes it quite clear that the concept "mystical" must not, and cannot, be confused by an objective reader with "mysticism." Rather, he already then maintains that a true science of nature needs a spiritual insight, a science of the spirit, to fully comprehend the phenomena that present themselves to humanity. In this sense, he describes in the book the mysteries of ancient times, particularly in Greek civilization, and their rebirth in the mysteries of the Christ Events.

With this course of lectures Rudolf Steiner set an example of how a modern humanity can respond freely to the challenges that occur in connection with the return, for instance, of the planets to the positions at the time of the inauguration of

Christianity. He set similar examples on many other occasions, but this one is particularly significant because it stands out like a guide with regard to the perspectives of the Second Coming, which we are discussing right now.

On a later occasion during Rudolf Steiner's lifetime, in 1915–1918, Saturn moved into the actual positions that it took up during the Three Years, that is, it moved then through the ascending node of Jupiter, its own ascending node, and so forth. Again he set an unprecedented example, or rather, a road sign for a humanity on the threshold of the Second Coming. During those years—it was the time of the First World War—he carved and worked at his "Statue of the Representative of Humanity." This statue exists and can be seen in the Goetheanum, the central building of the anthroposophical movement, in Dornach, Switzerland. We know that it represents the Christ as Rudolf Steiner perceived Him by his clairvoyant, spiritual insight. It shows the Christ in a situation that is particularly significant for the human being of the present age. His left arm and hand are lifted up, as it were, to cosmic heights. The gesture is rather one of bringing down those heights to Earth reality. The right arm is directed toward the Earth below, in a protective attitude. Above in the heights we see Lucifer, falling down, as it were, from realms of earth-estranged cosmic being where he lives in immeasurable pride and rejection. He is one of the two adversaries against whom we must exercise determined recognition and our own power of protection. Below the Christ, in depths of subterranean caverns, Ahriman, the other adversary, is bound. It is he who attempts to fetter us to the material world, and to its exclusive recognition.

Of course, these adversaries were not idle either at such times when the living memory of the Christ Deed was activated by the cosmic occurrences into the realization of His Presence. For them it is a matter of the frustration of their intentions with regard to the future of humanity. And their reactions we can well recognize in history. For instance, just toward the end of the First World War, when Saturn moved through the ascending node of Neptune in 1917–18, there was such an occasion. Rudolf Steiner himself was then urged to respond to human anxieties in his environment with regard to the burning social problems that arose, particularly in Central Europe. He suggested then the "Idea of the Threefold Commonwealth," by which he gave far-reaching details with regard to healthy solutions concerning all spheres of human community life. Had these ideas been realized, they would have inaugurated a true, practical Christianity, right down into concrete details of our earthly existence and without any sentimentality. As soon as it was pronounced, it was combated by the cohorts of the adversaries with all possible might. At the same time there was created an active counterstroke in the social sphere. That was the introduction of Leninist communism into Russia (Nov. 9, 1917). Built into dialectical materialism was the idea that the human ego had to be eliminated, because it was the culprit that created the social crises. And the elimination of the ego is just the impulse that the anti-Christian forces perpetrate. In Christ the greater, the cosmic, Ego appeared, which alone can lift us out of the narrowness and destructiveness of the small ego and egotism. This, however, the adversaries cannot accept for the sake of their own existence.

When the thirties of this century approached—and with them the year 1935, the commencement of the Second Coming—the adversaries became particularly active. They spoke and worked through people like Hitler and his associates. Through them they promised work and bread to a desperate humanity in Central Europe, but at the price of discarding the ideas and impulses of a true Christianity. In that moment, during the first half of 1933, Saturn moved through the descending node of Neptune, opposite the configurations of 33 A.D. Before that, the adversary forces, one could say, prepared this situation very carefully. With the crisis in the world economy, initiated by the events in October 1929, they tried to throw a weakened humanity back into the bondage of matter and its preponderance over anything of a spiritual nature.

The following events in world history show how these forces use the instrumentality of apparent external pressure and fear to try to achieve their ends. One of the results of the take over in Central Europe by Hitler and his associates was World War II. It started in 1939, and even with all the fierceness by which it was conducted, dragged on endlessly. Eventually, on August

6, 1945, as we know, the U.S.A. used the atom bomb at Hiroshima. Since that time humanity has been living in a kind of perpetual anxiety about the cataclysmic perspectives that then opened up. And fear is one of the prominent whips that the adversaries use, apart from promises, to hold humanity down and to frustrate any traces of world hope in the sense of a genuine Christianity. On that date, Saturn had returned to a position between the ascending node of Jupiter and its own ascending node, as in 31–33 A.D. Apart from this, Pluto was in the ascending node of Neptune.

We have, so far, been speaking almost solely of the transits of Saturn. However, we must realize that all the other planets also return in rhythmic cycles to the original configurations. It would lead us too far to discuss all these details. But we can be certain that all these events also work as challenges into the life of modern humanity in connection with the Second Coming. They are also connected with the experiences of individuals in this context, though they are mostly kept in the secrecy of personal soul life. We have, though, a good number of reports of such experiences, particularly during those years beginning with the thirties and right through the war. So, the Second Coming is not just a fictional story of what some people may be inclined to attribute to religious mystics, but it is a sphere of concrete experiences and events. The counterattacks of the adversaries draw attention away from their reality. Yet they have by no means won the "Great Battle," which can give us justified hope.

The facts presented in the foregoing description can give us the confidence that the study of the stars in connection with the life of Christ is not just a matter of curiosity. It brings with it deep and vital implications with regard to the very existence of humanity in a historic sense, particularly concerning the present age. The connection of Christ with the cosmos that Rudolf Steiner mentions in *The Spiritual Guidance of Mankind* (Lecture III)—"He [the Christ] made no step without this working of the cosmic forces into and in Him.... The forces active within Him, however, were cosmic forces, coming from the Sun and stars"—is of greatest significance, even beyond the mere "Three Years" during which it happened. We can, for instance, ask: Why did John the Divine write the Book of Revelation, the story of the Apocalypse? He did it because this is an essential part of the very Being of Christ, of Him of Whom John says (Revelation, 1: 7–8): "Behold he cometh with clouds; and every eye shall see him, and they also which pierced him... I am Alpha and Omega, the beginning and the ending saith the Lord."

In this sense it has also revealed itself, in the course of many years of research, that the Presence of the risen Christ will not only make itself apparent in connection with the so-called contemporary transits of the planets that we described above. For in the very configuration of the stars during the actual Three Years, the past and the future of the present universe is contained. In the sense that Christ identified Himself, in His immeasurable love, with those configurations, He is the Being of the Apocalypse. There are a number of ways to demonstrate this. One way is to again take Saturn, the organ of cosmic memory, and work with it according to a certain transmutation of time, similar to that which we employed for the mysteries of the Second Coming. Saturn can know (he is representative, though only representative) the will of the heavenly Father. In the Acts of the Apostles, 1:7, we hear: "It is not for you [as human beings] to know the times or the seasons, which the Father hath put in his own power." We take one degree of the movement of Saturn in the zodiac as representing one Saturn "year," that is 29.4577 Earth or Sun years. Thus we find an "Apocalypse" of the Earth and the universe. To illustrate, a movement of Saturn 67.5 degrees forward from its heliocentric position at Golgotha brings us to 180 degrees. This can be converted into the equivalent of the present age and the near future, i.e., to 2021 A.D. (67.5 X 29.4577 = 1,988 years, plus 33.35 years = 2021 A.D.) We have chosen this deliberately to illustrate a remarkable incident. This alone, and we could add to it all the positions of the other planets, describes our present, and the near future, in perfect apocalyptic terms. It tells us that we have entered an era of "weighing," of decision in the highest sense. Because Saturn in 180 degrees relates itself to that part of the ecliptic in which, in the geocentric perspective, the Sun appears at the commencement of autumn, in the "sign" of Libra. A careful study of all the problems and vistas that present humanity has created can give us the vivid impression that we are moving fast toward a moment in history in which the fundamental spiritual and ethical principles of being human will have to be reviewed and corresponding decisions taken on this basis. Many have already come to the conclusion that this present humanity is slowly drifting toward the edge of abysmal self-destruction. There is a certain truth in this, and fundamental changes are required to avoid it. Therefore the radical reappraisal of all human values, both moral and also in very practical connections, has to be effected. All this is indicated in Saturn's "progression" of 67.5 degrees from its position on April 3, 33 A.D. to 180 degrees of the ecliptic.

We must ask: What can be the means, so to speak, the objective gauge, for a decision of such major dimensions? Here, the Star-Apocalypse in connection with Golgotha can give an answer too. Earlier, before Saturn was in 180 degrees, it had a conjunction with Uranus at about 158.5 degrees. This was about 40 degrees beyond the position on Good Friday, 33 A.D. This brings us to the time around 1380 A.D. Before this happened, Saturn was in the place where Uranus actually stood on April 3, 33,—28 degrees forward from the heliocentric location of Saturn at Golgotha. This refers us to the middle of the ninth century A.D., or 858 A.D. (28 x 29.4577 = 824.8 yrs. + 33.25 yrs. = 858.25) (140.7 degrees [Uranus] minus 112.5 degrees [Saturn in April 3, 33] = 28 degrees).

It is now a matter of comprehending the historic meaning of these dates. During the ninth century an event occurred that is of archetypal significance for humanity of the present age. The personality who is known to us only through the tradition of the poetry and literature on the Holy Grail, Parzival, is described to us as a human being who at one time beholds the Holy Grail, the most sacred presentation of the higher meaning of "humanhood." However, he does not recognize its significance, and consequently has to go through a long spell of outer and inner battles, of despair and utter loneliness. Finally, he is again led by circumstances into the presence of the Holy Grail. Now, after his trials, he can comprehend its significance and can even become King of the Holy Grail. The road that Parzival had to take, his long preparation and catharsis, can become for present humanity something like a shining archetype, a message of hope, above all religions and denominations.

This great beacon of human hope was then transformed in the Middle Ages. Around 1380 a human being entered the world who then became known under the name Christian Rosenkreutz. The movement that he inaugurated, and particularly the uniting "sign" of the Rose Cross, are expressions of the earlier Grail motif, in a form appropriate to that age. The black cross that carries the roses, the symbols of life, is a transformed image of the Holy Grail. It was to be the guiding beacon for the pupils of Christian Rosenkreutz, who had resolved to redeem humanity from the descent of evolution into material existence, and to put thereby the "house of Man" on Earth in order. Thus the "progressions" of Saturn in connection with the stars of the Three Years reflect the "Apocalypse" of later ages. On the one hand we read in it that present humanity is moving quickly, motivated by contemporary events and catastrophes, toward great decisions. On the other hand, we can also discern in it the prerequisites, as it were, that can lead us to recovery and redemption. It is simply a new spiritual comprehension of the being of our humanness, where we come from, where we go to, what the meaning of our existence is. Of course, all this is tied up with an understanding of the evolution of the whole universe in which we live. In other words, we need a science of the spirit in order to cope with the burning problems of our age. The Grail Movement and Rosicrucianism brought this "science" to their contemporaries, in a language that they could comprehend. Our present age saw the endeavors of Rudolf Steiner, who did not discard those earlier messages but who translated them through his "Science of the Spirit," or Anthroposophy, into a language that modern humanity can comprehend. He regarded, for instance, his work *Occult Science* as a science of the Holy Grail, and Anthroposophy altogether as a presentation of genuine Rosicrucianism in a form congenial to the mind of modern humanity. (See his *Theosophy of the Rosicrucian.*)

Constructive solutions in this sense would have a significance not only for the present age. Here, too, the Saturn "progressions" of the living cosmic Apocalypse can inform us unmistakenly. After having traversed 247.5 degrees of the ecliptic, from its position at Golgotha, 112.4 degrees, Saturn moved into the place where the Sun appears to be at the commencement of spring, 360 degrees or zero degrees. This is opposite the position referred to earlier as the present Age of Decision. This leads us to c. 7350 A.D. (247.5 x 29.4577 years), which is close to the termination of the present Fifth Post-Atlantean Epoch, that is, the cycle of seven civilizations, starting with the Ancient Indian, after the destruction of Atlantis. That time, the eighth millennium A.D., will again see unimaginable destruction, the man-made termination of this whole epoch, just as Atlantis was destroyed by nature catastrophes. On the other hand, a part of humanity, however small, will have to effect an "Exodus" in order to reinstate the course of human evolution, under whatever circumstances. This is already present in the contemporary great crisis demanding decision. The solutions, if they are found and accepted now, will affect that distant future.

BIBLIOGRAPHY

Publications by Willi Sucher

Isis Sophia (1951–1952)
Ancient mythologies of the constellations of the zodiac, stages of cosmic
evolution and the creation of the solar system.

Man and the Stars (1952)
Includes a build-up of the relationship of the human being to the world of
the stars at the moment of death, embryology and the stars, physiological
deformities and the stars, the cosmic and physiological foundations of soul
life, and cosmic rhythms and the evolution of humanity.

The Drama of the Universe: The New Interpretation (1958)
A basic book on astrosophy in which the foundation of the modern heliocentric approach is developed. The character of the planets is discussed
together with the fixed stars and their mythologies. Great emphasis is
given to the correlation between the planets and historic events in nature,
among nations and with individuals.

Star Journals (1965–1970)
Published here are the first 14 issues from the original 61, which outline
the essential properties of the zodiac with the surrounding fixed stars and
the planets of the solar system.

Practical Approach Towards a New Astrosophy (1974)
Practical details of how a star configuration can be worked out. Introduction of the heliocentric approach in great detail.

All of the above are published by the Astrosophy Research Center,
PO Box 13, Meadow Vista, CA 95722

BIBLIOGRAPHY

Basic works by Rudolf Steiner

Occult Science

Knowledge of the Higher Worlds and its Attainment

Theosophy

The Philosophy of Spiritual Activity

Some recommended titles from the works of Rudolf Steiner related to astrosophy

The Spiritual Guidance of Humanity

Man: Hieroglyph of the Universe

Human and Cosmic Thought

Spiritual Beings in the Heavenly Bodies and in the Kingdoms of Nature

The Spiritual Hierarchies and their Reflection in the Physical World

Supersensible Man

Between Death and Rebirth

The Evolution of the Earth and Man and the Influence of the Stars

The Influence of Spiritual Beings Upon Man

Man in the Light of Occultism, Theosophy, and Philosophy

Man and the World of Stars

For information on the availability of the above titles by Rudolf Steiner, please write: Anthroposophic Press, RR4, Box 94 A-1, Hudson, New York 12534, U. S. A.

WILLI SUCHER

in the late 1950s or early 1960s

(Courtesy of the Astrosophy Research Center)